W0113585

FISHING EUROPE'S TROUBLED WATERS

Spanning the last 50 years of fisheries policy in Europe, this book is the parting con-tribution and career-spanning reflection from one of Europe's most renowned social scientists working in the field of fisheries management and policy.

The last 50 years have without doubt been the most turbulent years in the history of North Atlantic fisheries – a turbulence brought about by the actions of fishers, scientists and above all politicians. It is a period of change that sees a radical redraw-ing of the political geography of fisheries, globalisation of trade, the development of fisheries management towards increasingly restrictive regulation, and declining fish stocks. The book explains why the bold but deeply flawed Common Fisheries Policy persistently failed to deliver its basic goal of sustainable fisheries. The spotlight falls on the monolithic, highly centralised, command and control nature of the Policy that strives to apply a universal 'one size fits all' approach, thus creating a governing system wholly unsuited to the system to be governed, out of kilter with preferred models of governance, and disconnected from the practical realities of fishing as a livelihood in a challenging environment. A final section on Brexit focuses on its halting progress from concept to reality, the implications for the fisheries sector and the fateful final negotiations with the EU over the fisheries question. Seeking to explain why the anticipated benefits for the UK industry failed to materialise, attention is drawn to the misplaced political hubris over regaining 'sovereignty' in areas like the North Sea.

This book will be essential reading for students, scholars, professionals and poli-cymakers working on fisheries, marine governance, natural resource management, environmental policy and the European Project.

David Symes, Reader Emeritus in Geography at the University of Hull, was one of Europe's most renowned social scientists working in the field of fisheries man-agement, rural development and policy. During his academic career, which spanned seven decades, he generated a prolific volume of research and had a major impact on fisheries policy and the formation of a European fisheries social science community.

Earthscan Oceans

For more information about this series, please visit: www.routledge.com/
Earthscan-Oceans/book-series/ECOCE

FISHING EUROPE'S TROUBLED WATERS

Fifty Years of Fisheries Policy

David Symes

LONDON AND NEW YORK   from Routledge

Designed cover image: © Getty Images

First published 2023
by Routledge
4 Park Square, Milton Park, Abingdon, Oxon OX14 4RN

and by Routledge
605 Third Avenue, New York, NY 10158

Routledge is an imprint of the Taylor & Francis Group, an informa business

© 2023 David Symes

British Library Cataloguing-in-Publication Data
A catalogue record for this book is available from the British Library

Library of Congress Cataloging-in-Publication Data
Names: Symes, D. G. (David Gilyard), author.
Title: Fishing Europe's troubled waters : fifty years of fisheries policy /
David Symes.
Description: New York, NY : Routledge, 2023. | Includes
bibliographical references and index.
Identifiers: LCCN 2022044414 (print) | LCCN 2022044415 (ebook)
Subjects: LCSH: Fishery law and legislation—Europe—History. |
Fishery law and legislation—European Union countries—History. |
Fishery law and legislation—Great Britain—History.
Classification: LCC SH253 .S95 2023 (print) | LCC SH253 (ebook) |
DDC 639.2094—dc23/eng/20220928
LC record available at https://lccn.loc.gov/2022044414
LC ebook record available at https://lccn.loc.gov/2022044415

ISBN: 978-1-032-42471-2 (hbk)
ISBN: 978-1-032-42470-5 (pbk)
ISBN: 978-1-003-36291-3 (ebk)

DOI: 10.4324/9781003362913

Typeset in Bembo
by codeMantra

David Symes, B.Litt. M.A. (Oxon)
Reader Emeritus in Geography, University of Hull
Born 30 July 1934, Died 13 January 2022

CONTENTS

ABOUT THE AUTHOR

David Symes, Reader Emeritus in Geography at the University of Hull, had an academic career which spanned seven decades, over which he generated a prolific volume of research and had a major impact on fisheries policy and the formation of a European fisheries social science community.

Born in Bradford, at school he excelled academically. He was partially self-taught in Geography and secured a King Charles 1 Scholarship at Jesus College Oxford, where he was awarded the Herbertson Memorial Prize. Graduating in 1956 with First Class Honours in Geography and following a period of postgraduate research on rural change in western Norway while based at the Norwegian School of Economics in Bergen, he was appointed to a teaching post at the University of Hull in 1958, where he would work in the Geography department for over 40 years up to 'retirement' in 1999.

For the first 25 years his research focused on the social structures and dynamics of family farming systems in Britain, Ireland and Europe, when he would also develop his fascination with and understanding of North Atlantic fisheries. Working alongside sociologists from western and eastern Europe on rural communities and families in industrial societies, and involvement with the European Society for Rural Sociology as Scientific Secretary, Vice President and Co-editor of *Sociologia Ruralis*, he became a leading light in rural development.

From the mid-1980s onwards, David increasingly focused his research on fisheries and their management, stimulated by issues relating to supply and distribution within the UK and emerging problems confronting the nascent Common Fisheries Policy, which continued well beyond retirement.

During the 1990s he coordinated two major EU-funded projects, the first on Devolved and Regional Management in Fisheries (1993–1995), which would provide formative evidence for successive reforms and efforts to regionalise the Common Fisheries Policy. The second, a Concerted Action, established a European Social Science Fisheries Network (1996–2000). 'ESSFiN' helped to change the course of fisheries social science and the approach to fisheries management, bringing new insights into institutional change, the role of applied social science and multi-disciplinarity, inter alia. Under David's leadership ESSFiN successfully initiated and developed an active network of social scientists – very broadly defined – with the aim of deepening the role of socioeconomic issues in fisheries governance. It was a prolific project, producing six major edited volumes setting out the extent and breadth of the social sciences' potential in relation to fisheries governance and management: *Property Rights and Regulatory Systems in Fisheries* (1998); *Northern Waters: Management Issues and Practice* (1998); *Europe's Southern Waters: Management Issues and Practice* (1999); *Alternative Management Systems for Fisheries* (1999); *Fisheries Dependent Regions* (2000); and (edited with Jeremy Phillipson) *Inshore Fisheries Management* (2001).

ESSFiN was, and still remains, a unique experiment in applying a multi-disciplinary approach to fisheries management. It demonstrated the relevance of the social sciences to an understanding of current issues in the organisation, development and governance of marine fisheries in Europe. As David explained himself: "Fisheries management at the time lacked a coherent body of evidence describing how the regulatory regime was not only undermining the social structures and value systems but also turning fishers into the objects rather than the subjects of fisheries policy engendering low commitment and weak compliance on the part of the fishing industry". It would become a standard bearer for an applied social approach to fisheries governance and management.

David would develop valued and longstanding relations within the University of Hull, notably with colleagues in the Hull International Fisheries Institute, the Institute of Estuarine and Coastal Studies and the European Studies department. He garnered huge respect nationally and internationally across the worlds of academia (in both the social and natural sciences), policy and the fishing industry. In 2021 he was awarded an Honorary Fellowship at the Centre for Rural Economy, Newcastle University.

David carried out many research contracts from the Sea Fish Industry Authority, Joint Nature Conservation Committee and English Nature. Following retirement, in 2000 he was appointed as the prestigious Buckland Professor. He served on panels of inquiry into the future of the fishing industry, established by the Cabinet Office Strategy Unit, the Royal Society of Edinburgh and the Scottish Government, and on the editorial board of Fisheries Research (2000–2010). In 2007, he was invited by the European Commission to critically reflect on and make recommendations for reform of the Common Fisheries Policy, thus kick starting the decennial review process and helping shape proposals for the 2012 reform.

David would again re-emerge from retirement in 2017, dismayed by the outcome of the 2016 referendum and provoked by the realities of leaving the EU for the fishing industry, fisheries management and the seafood supply chain. As he had done throughout his career, he took his pen to paper to embark on a major reflective book on the last 50 years of fisheries policy in Europe.

FOREWORD

I first met David Symes at the beginning of the 1990s as a student reading Geography at Hull University and attending his highly stimulating and widely appreciated lectures on sustainable fisheries management. Shortly after graduating in the summer of 1993, I stumbled across a *Guardian* job advert for a Research Assistant post on his new EU-funded project 'Devolved and Regional Management Systems in Fisheries'. I was easily hooked and jumped at the opportunity to apply. I could never have imagined then that we would go on to be lifelong friends and collaborators.

Skip forward a few decades and we would regularly find ourselves catching up in York railway station hotel enjoying half a pint of Yorkshire beer, chewing over developments in the world of fisheries and planning our next writing project. In the lead up to and after the referendum in 2016, our preoccupation had been Brexit and its implications for UK and EU fisheries. At one such meeting early in 2019, as the rancorous debates over the Withdrawal Agreement were in full swing within the UK House of Commons, David would first moot his intentions to produce a monograph. '*The Book*', as it came to be known between us, was to be a personal, programmatic reflection on the evolution of European fisheries policy, which would also set the scene and to some extent explain the relationship between Brexit and the vexed fisheries question.

Fishing Europe's Troubled Waters is the first book to span the 50 turbulent years in the history of Europe's fisheries and fisheries policy. It offers reflections on the changing circumstances and problems that have shaped Europe's fisheries and their management, beginning with the loss of distant water fishing opportunities in the 1970s and the negotiation and development of the Common Fisheries Policy (CFP) and ending with the UK's withdrawal from the EU in 2021.

David's reflections are grounded in social science research in UK and European fisheries over the 50-year period, conditioned by his strong commitment to the European project and tempered by a balanced though critical view of the CFP's performance and the consequences of Brexit. We learn through its pages just why the "bold but flawed CFP", while successfully negotiating a stable, long-term allocation of fishing rights within the 'common pond', has persistently fallen short in delivering its basic goal of sustainable fisheries. The spotlight falls on the profound mismatch between a highly centralised and inflexible governing system that strives to apply a universal 'one size fits all' approach, inimical to radical reform, and what is an extensive, diverse and complex 'system to be governed' – a policy out of kilter with preferred models of participative and interactive governance built on an appreciation of the behaviours, values and aspirations of fishers and other actors.

David was always convinced that effective fisheries governance in semi-enclosed seas, with a prevalence of shared stocks, requires integration rather than segregation, collaboration rather than isolation, within the context of a regionalised approach, which he saw as the "key to securing a truly radical reform of the CFP". Indeed, despite its flaws, and his often hard-hitting criticism, David still saw the CFP as "Europe's fisheries best hope for the future" and that the EU is better off with it than without it. Readers of the book will therefore quickly discern his dismay by Brexit and its implications for fisheries and the livelihoods they support, arriving, as it did, at a time when the CFP was making progress towards realising its ambition and becoming a "half decent international policy instrument". He gives us an account of Brexit's halting progress from concept to reality, the consequences for the fisheries sector and seafood supply chains and the fateful final negotiations with the EU over the fisheries question when, to all intents and purposes, the rhetorical claims and anticipated benefits for the UK industry failed to materialise, leaving it to face an uncertain future and jeopardising a joint vision and collaborative management for seas and resources shared with the EU.

Fishing Europe's Troubled Waters provides a 50-year retrospective from 1970 to 2020 – it equips readers with the keys to understanding how it is we came to where we are now in European fisheries policy. Yet, the book is also intensely forward-looking, setting out the prevailing uncertainties and challenges that lie ahead and the necessary ways forward for securing viable and sustainable fisheries. It is vital reading for those responsible for putting in place, or simply interested in, the essential coordinates for effective fisheries governance in the face of environmental, economic and political instability. As David sets out, in typically uplifting and programmatic terms:

> What we can do is begin the much-needed transformation of fisheries management, from its deeply embedded attachment to 20th-century modes of thinking in both the science and management of fisheries structured around equilibrium models that have led to the creation of simplistic, reiterative management plans, to a way of thinking that embraces uncertainty and the threats of potentially catastrophic disequilibrium. Both marine

environmental and fisheries management will need sooner or later to abandon their preservation strategies and replace them with a management approach that builds in the need for greater flexibility, resilience and adaptive responses. There is no better time than the present to start the process that will improve performance now and prepare the industry for the challenges of the future.

By May 2019, typically expedient, David had produced a full draft of the introductory chapter to the book. At this stage he knew clearly what he was envisaging as the end product, explaining that "All I'm looking for at this stage is a broad comment on the general style (attempting to move away from that of the journal article to provide access for a wider readership)". This was not going to be a conventional, tightly referenced, academic book, but rather one which would have wider accessibility and attract not only academics, students and those directly involved in fisheries and their management, but also a wider readership interested in the history of fisheries and the European project. David would express some worries as to how this style would sit with academic publishers, but wanted to pursue the project anyway even if it eventually meant an in-house publication. His motivation was to set out his own personal reflection on fisheries policy and what, in all likelihood, would be his last major work.

David would go on to produce draft chapters out of order sequence. By November he was working on 'Brexit and beyond' as the "concluding section of my personal take on the events surrounding fisheries policy in Europe over the last 50 years". In April 2021 David described that he had "now completed the text, having slightly revised its structure to permit a separate chapter on the events of 2020. This is currently being typed (by A!) and I hope to let you see the results in a couple of weeks' time". In September that year I would drive to meet David at his home in Beverley to pick up the bulk of the handwritten manuscript which we agreed was too precious to risk in the post. Our first in-person meeting in almost two years due to COVID-19 would turn out to be our last face-to-face catch up. We discussed the book proposal and an exploratory message I'd left with Routledge to flag up the interest, which would lead to discussions with the publisher and an early drafting between us of the formal book proposal in October. Minor corrections to specific chapters would follow, including to Chapters 1 and 2. On 22nd November, David wrote, in what would be his last email to me:

> Re the book, I am presently toying with adding a very short Chapter 11, reflecting on how much has actually changed over the last 50 years, especially in terms of the relationships between the governing system and the governed. When I've given this more thought, I will get back to you with more details.

David Symes died on 13th January 2022, in his 88th year, with his handwritten manuscript tantalisingly close to completion. Within an instant the book would take on added significance as the final contribution and career spanning perspective from one of Europe's most renowned social scientists working in the field of fisheries management, rural development and policy. As readers of *Fishing Europe's Troubled Waters* may come to find for themselves, David was one of a kind – progressive, always ahead of the curve, a true scholar and a critical thinker. Agile, articulate and sharp witted. Not afraid to speak his mind. Dedicated to his craft and for many a reference of academic excellence. A masterly and meticulous editor and author. Never esoteric. Always engaging. Often mischievous.

Though the manuscript was almost complete at the time of his death, finalising it has involved a number of steps and judgment calls that should be flagged. It should first be acknowledged that, although David had provided revisions and edits to individual chapters, he did not have the opportunity to stand back and make final alterations to the overall manuscript. David would have undoubtedly undertaken this step and introduced some final finessing. However, readers will find that the book holds together very well, which is a tribute to David's writing skill and ability to identify and follow a narrative, together with his writing practices that relied on handwritten manuscripts and revisions.

David was very careful and precise with his use of words, and firm in his intent. He would not have welcomed substantial interference with his writing. I have therefore resisted making any changes to the text or its arrangement, beyond some necessary, minor corrections. Some small inconsistencies in tense have also been addressed, which arose from the timing of the production of specific chapters in relation to the Brexit timeline – thus occasional references in earlier chapters speculating over the outcome of Brexit negotiations have been removed, accepting that in later chapters David discusses these initial outcomes. Chapter 8 on Brexit and its implications required some slightly heavier editing to tackle this issue. Discussions with David's wife Annick, who was closely involved throughout the writing of the book, suggest that we can be almost, though not entirely, sure that we have included 'final' versions of Chapters 9 and 10. However, although they had already undergone revision, it is possible David would have revisited both these chapters after completing his additional Chapter 11.

Chapter 11 has been presented in its unfinished state without alteration or elaboration. It was a great relief to find that a full neat manuscript of the chapter had been located in David's file. In it, in the final words that David would put to paper, he fittingly returns full circle to themes that have engaged him since the very beginning of his career around small-scale coastal fisheries and the social dynamics of the fisher household, family and community. This draft version of the chapter is included in the book, though we know it was a work in progress at the time of his death. He had also intimated previously to Annick that he was stuck over the conclusion, and we don't know whether he had resolved the dilemma. David would almost certainly have added a concluding remark to the chapter to end the book on a wider, programmatic note. With the chapter itself being a late decision, readers might wish to look to the original end statement of the book at the conclusion of Chapter 10 and David's discussion of climate change. The Chapter 11 manuscript was also accompanied by a folder containing further handwritten jottings and notes that may (or may not) have been woven into the final chapter. With the help of a magnifying glass, and aided by years of collective experience between Annick and me in reading David's neat but deceivingly difficult-to-decipher handwriting, these unincorporated jottings are presented in a final section of the chapter, with only light editing. They provide a small window into David's writing process.

Two final considerations require brief mention. First, maps have been designed, drawn and located in the text posthumously, according to a list of Figures David had passed on to me and some brief guidance as to what they should contain. David was a great supporter of including maps in his publications, so rather than take a decision to leave them out, his brief guidance has been used to prepare what we think he had in mind. He would almost certainly have fed back constructive comments to improve the maps we have produced for the book. Any errors contained within them are my own.

Second, and perhaps most significantly, following discussions between David's family, myself and the publisher, the book's title has been revised. Revision follows recommendations by the publisher to enhance the book's searchability and avoid confusion with other titles. The original title for the book that David envisaged was: *Troubled Waters: Reflections on European Fisheries Policy 1970–2020*. We have held to his original as closely as possible.

Finalising *The Book*, and engaging with David's final piece of writing, has been a poignant and personally challenging experience. Over the months I have often imagined hearing David's words of encouragement ringing in my ears. Now it is complete there are mixed feelings about bringing to a close the final chapter of a collaboration that has spanned 30 years. I am especially grateful to the enduring support of David's family, his wife Annick, daughter Janet and son Nigel, who were also pivotal in the final production of the book. Also for the good wishes from the scores of fisheries social scientists and others across Europe who gave encouragement and offered help if it was needed to see David's last work over the line. Thanks are also extended to Karen Stubbs at Newcastle University who in her spare time transposed many of the handwritten chapters. My appreciation also goes to our design team at *The Works* for their production of the maps. Finally, David's family and I are indebted to our Routledge editor Hannah Ferguson for encouraging and seeing the importance of the project and in bringing to fruition David's parting gift to us all.

Jeremy Phillipson,
Professor of Rural Development,
Newcastle University, UK

PREFACE

I am sometimes asked how I came to be involved with fisheries. My response is usually built around two place names: Norway, that sparked my interest, and Hull, that sustained it. In 1957 postgraduate research took me to western Norway where my intentions were to explore the processes of modernisation in two contrasting but equally challenging environments: the coastal fringe and the inner fiords. Although my focus was primarily on rural, largely agricultural communities, it was impossible to ignore the role played by small-scale, highly seasonal coastal fisheries that provided an important supplementary income for many in the island communities. Over the next few years, I regularly returned to the coastal fringe, eventually moving north to islands where fishing was a mainstay of the local economy. I was already well and truly hooked.

In 1958 I had taken up my first appointment as an assistant lecturer in geography at the University of Hull. At the time Hull was the major distant water fishing port in Europe. Again, it was impossible to avoid the huge influence that fishing exerted on the city and the shock of losing its access to those distant waters in the 1970s. Over the next two decades my interest in, and understanding of, fisheries was broadened by fieldwork in other parts of the Atlantic fringe including Ireland, the Færoes, Iceland and Newfoundland and deepened by work commissioned by Seafish relating to the seafood supply chain. During this time, too, my academic perspectives as a social scientist and Europeanist were much influenced by working alongside European partners in a UNESCO-funded project on the future of the rural community in industrial societies spanning both western and eastern Europe, led by the renowned French rural sociologist, Henri Mendras.

It was not until the late 1980s that I took the decision to quit the rather overcrowded field of rural studies to focus my research exclusively on fisheries in general and the EU's bold but flawed common fisheries policy in particular. There followed the most active and rewarding period of my career, triggered by a series of EU research grants relating to the social dimension of fisheries management and involving partners in Norway, Denmark, the Netherlands, France, Spain and Greece, and by work undertaken for statutory nature bodies in England, Scotland and Wales. These commitments took me up to, and in some cases beyond, retirement in 1999 from my teaching post in Hull. I had no intention of giving it all up, vowed to slow down and enjoy a more leisurely approach to fisheries. In 2007 I was rewarded with an invitation from the European Commission to undertake, together with Mike Sissenwine, a critical analysis of the CFP as a preparatory step for the next decennial review. Although this involved a brief 'headbanging' exercise it proved a very satisfying experience and potentially an appropriate 'swansong'. Seven years on I again decided to quit the scene gradually and quietly!

So what caused me to reengage with fisheries policy in 2018/2019? It was, of course, Brexit. As a Europeanist, I was naturally dismayed by the outcome of the 2016 referendum, though content to let things take their course. However, I became increasingly concerned over the handling of Brexit, the sweeping denunciation of the CFP and the future of fisheries in the UK and Europe. I was made to reflect on the changes to fisheries and fisheries policy over the past 50 years that covered the end of the *ancien régime* of fisheries as an open access common property resource, the setting up of the CFP and its slow, unsteady progress towards becoming a half decent international policy instrument prior to the unexpected fracturing of the UK's relationship with Europe that was bound to have serious consequences for fisheries management in the seas around Britain's coasts.

Those reflections are contained within this volume. It is not intended as a detailed academic account of events and the text is thus largely devoid of citing sources; where certain texts have been extensively used to add substance to my understanding of what was going on, they are noted at the end of each chapter. The hope is that the book might reach a far wider audience than is found in the academic world.

I owe a huge debt to the very many people who have helped in different ways to guide and develop my understanding of fisheries and their management, be they fishers, merchants, processors, managers, politicians or academic colleagues across a wide sweep of the North Atlantic basin from Kings Cove in Newfoundland to the many fishing communities, fisheries institutes and government offices within Europe. They are too numerous to name, and it would be invidious to single out particular persons. I will, however, mention two people who have been a source of help and encouragement over a long period of time and instrumental in the preparation of this book. The first is Jeremy Phillipson, an excellent foil to my sometimes bizarre opinions, a fantastic manager of research projects both past and present and a constant friend. The other is my long-suffering wife, Annick, for a lifetime of tolerating my obvious obsession, her diligence in seeing to all my IT needs and her forbearance concerning my occasional neglect of domestic duties.

NOTE FROM ANNICK SYMES

Further to David's appreciation in his preface, may I, on behalf of myself and David's family, add our sincere thanks to Jeremy Phillipson who, on David's untimely death, so willingly and capably stepped in to rescue the book. It is through Jeremy's hard work and dedication – with the encouragement of the Editor, Hannah Ferguson – that the book has been brought through the process of completion and publication in a manner that we believe David would have found eminently to his liking.

Annick Symes

1

INTRODUCTION

Introduction

As almost anyone who has taken a seaside holiday will be aware, the sea is never still. It is the incessant change in mood and motion that is the very essence of its enduring fascination. For all of the past 50 years the underlying mechanisms that govern the everchanging physical state of our seas and oceans – the predictable tidal regimes, the circulation of ocean waters and the pattern of currents that underpin the fertility of marine ecosystems, together with the rapidly changing weather conditions that can profoundly alter the state of the surface waters – have been well understood by scientists and their consequences all too familiar to those whose lives and livelihoods depend on the sea. Nonetheless, managing the marine environment and its resources remains a major challenge.

During the early part of the period, scientists felt safe in assuming that the ocean system was a stable one – one that witnessed both short-term variation and periodic cyclic fluctuation but was held in a long-term equilibrium. Over the past 30 years this comforting sense of stability has been rudely challenged by the evidence of global climate change. Marine systems bear the immediate brunt of the ensuing dislocation as melting polar icecaps release vast quantities of cold, fresh meltwater into the adjacent seas while elsewhere the surface waters of the oceans are rising. The pace of meltdown in high latitudes has surprised and alarmed the scientific world, fearing that human populations may be too slow in reacting to the challenge. Most political leaders have, in fact, been quick to pledge support for long-term strategies to reduce CO_2 emissions in the hope of slowing or even halting the process of global warming. But so far few have been as keen to introduce the necessary but costly support measures needed to secure the distant goals.

DOI: 10.4324/9781003362913-1

While we may need to be reminded from time to time of the consequences of the changing physical processes governing the state of our seas the focus of this volume is on a different kind of turbulence that affects the ways in which society perceives, treats and seeks to manage the living resources of the sea. The sources of that turbulence include the behaviour of fishers with a growing capacity for harvesting the seas' living resources, changes in international law governing the allocation of fishing rights, the quest for sustainability and the actions of policy makers operating at different spatial levels and struggling to establish fair and proportionate systems of resource management in a rapidly changing political environment.

Sketching out the narrative

The years from 1970 to 2020 describe a period of unprecedented extreme turbulence for the fisheries of the North East Atlantic. What follows is a very brief summary of a more complicated story intended simply to set the scene: a more detailed account will unfold in subsequent chapters. The narrative is framed by two sequences of events, of differing magnitudes, that fundamentally alter (or threaten to alter) the circumstances under which fishing activity can take place and set in motion a new phase in the development of fisheries governance and management. The first of these, occurring in the 1970s, was a concatenation of two initially unrelated events that quickly became entangled: the seismic effects of UNCLOS III that transformed the political geography of the world's oceans and their fisheries and the first enlargement of the European Economic Community (EEC) from six to nine member states, the precursor to the establishment of a common policy framework for the fisheries of the North Sea and its neighbouring waters.

The changes to the United Nations Convention on the Law of the Sea that began to be enacted *de facto* in 1976 and were confirmed *de jure* in 1982 (UNCLOS III) legitimised the enclosure of the marine commons, bringing to an end the long-established principle of the high seas that had allowed the fishing industry free and open access to all waters beyond the narrow territorial waters claimed by almost all coastal states. In its place each coastal state was granted an exclusive economic zone (EEZ) extending out to 200 nautical miles (nm) or to a median line where the nominal EEZs of two member states overlapped. UNCLOS III conferred on the coastal states direct responsibility for managing the living resources of the sea within their zones. The impact on the fishing industry was immediate and considerable but varied quite markedly from country to country; as expected there were both winners and losers.

Significant changes were also taking place in the governance of western Europe. In the very early 1970s Denmark, Ireland, Norway and the UK were invited to apply for membership of the EEC and in 1973 access agreements were signed with three of the appellant states: only Norway decided against joining. The addition of Denmark, Ireland and the UK gave substance to the idea of an integrated approach to fisheries. Negotiations on a common fisheries policy began in 1976 on the prior assumption of the principle of equal access and were finally concluded six years later in 1982 with the first framework regulation governing the conduct of fisheries conservation policy for the EEC's exclusive fishing zone during the period 1983–1992. In effect this largely cancelled out the fragmenting effects of UNCLOS III with the merging of the EEZs of the seven Atlantic coastal member states into a single EEC fishing zone and allowing fishing vessels of the member states access to their traditional fishing grounds but under a common set of rules set down by the institutions of the EEC. The real impact of this would only become fully apparent in later years.

The decades that followed the inauguration of a common fisheries policy in 1982 were at best 'inglorious', involving a policy approach that was idealistic, simplistic and irrational in its original design, inflexible in the face of changing circumstances and necessitating a long and difficult learning curve for both policy makers and the fishing industry before it could begin to achieve its goals. It is easy to blame the EEC or EU for their early failings but in fairness the task facing the Brussels bureaucracy was huge: the more important commercial fisheries of the North East Atlantic had been overfished for some time; stocks were declining and despite greater fishing capacity catches were falling; and fisheries management was in its very early stages of development.

The pathways chosen by the European Commission involved minimum disruption to the preexisting patterns of fishing activity and placed an overwhelming reliance on relatively blunt output controls (catch quota) to regulate catch levels. The use of so-called technical conservation measures or TCMs – gear regulations, closed areas and seasons *inter alia* – with their potential for allowing greater discrimination within the management system was limited. Attempts to limit the growth of fishing capacity that lay at the heart of the overfishing problem were weak and largely ineffective. What became widely regarded as a simple 'one size fits all' policy approach failed to halt, let alone reverse, the depletion of the resource and despite the persistent evidence of its failure survived two further enlargements of the EEC in 1986 and 1995 that extended the 'common pond' southwards to include the waters off the Iberian peninsula and northwards into the Baltic Sea. Allocating blame for this failure is not easy and will be explored further in later chapters; at this stage let us say that all parties – fisheries science, policy makers and the fishing industry – must each share some of the blame. It was not until the very end of the century, and more vigorously in the first two decades of the 21st century, that the Common Fisheries Policy began to show convincing dividends with the return of some key stocks to 'within safe biological limits' offering the promise of realising the UN's Convention on Biological Diversity's 1992 target for fishing to achieve maximum sustainable yields (or MSY) by 2020. This was an achievement, reached through the persistence of the policy makers and the collaboration of the fishing industries, that was purchased at very considerable hardship and cost to Europe's fishing industry.

In marked contrast to the transformational changes of the 1970s, the events that were to challenge the governing structures of the previous 45 years may well be judged by future historians as 'a little local difficulty'. But the decision of the UK electorate in the referendum of 2016 to leave the EU 28 was significant in fisheries terms simply because of the pivotal geographical position occupied by the UK within the EU's 'common pond'. Ironically, the referendum took place at the very time when prosperity and confidence in the future of the fishing industry was being reasserted. The decision to leave had little or nothing to do with the fisheries question *per se*, even though the fishing industry was, almost to a man, strongly supportive of the leave campaign. Yet, the UK's departure from the EU and its CFP inevitably threatens the cohesion and stability of the systems of fisheries governance and management that were put in place during the late 1970s and early 1980s following the traumatic events of the early 1970s.

Aims

The purpose of this monograph is to provide a balanced, critical perspective on the last half century of fisheries governance and management in European waters. The circumstances surrounding the turbulence of the 1970s, the creation of the CFP and its early history of failure have been distorted, misunderstood and misrepresented over time – sometimes quite deliberately – in somewhat over-simplified accounts of both the origin of the upheaval and the subsequent role of the European Community/Union in its attempt to resolve the crisis in fisheries management in the 1980s and 1990s. Moreover, the realities of the most recent period, involving the stabilisation and partial recovery of the fish stocks and the resulting return of optimism over the future for some sectors of the industry – as reflected in rising levels of investment in new vessels and the modernisation of landing facilities – has been largely ignored. In its concluding sections, the monograph will focus on the pros and cons of future relationships between the UK and EU in the field of fisheries as a consequence of Brexit. The analysis will focus attention not only on the implications for the fishing industry *per se* but also on the wider concerns of the seafood supply chain and the impact on Europe's fishing-dependent communities. The overall analysis will aim to get closer to the truth of what went wrong, when and why, by looking deeper into the causes and consequences of the turbulent 1970s, examining the rapidly changing contexts in which fishing has been conducted over the last half-century and the increasingly strained relations between the principal actors involved in the unfolding story. Its intention, however, is not to allocate blame even though criticism will, from time to time, be directed at some of their actions or reactions.

The dramatis personae

Before commencing the narrative, it may prove helpful to identify and provide brief pen portraits for some of the actors involved. There are but two principal protagonists – the fishing industry and the policy makers – each with their own often contrasting world views, lifestyles and philosophies that seem to clash at almost every turn. Yet surprisingly both actors share the same purpose of securing the essential benefits of a renewable but vulnerable resource for both present and future generations. Where they differ is in the framing of the problem, the preferred evidence base and learning process and, perhaps most crucially, their choice of time scales. For the fisher it is the next trip, the next season and potentially the next generation. By contrast the policy maker will usually set their end goals in a longer and more elastic time scale, probably for at least ten years and maybe longer, though they may be called to account for their actions after only a very few years.

Key players: the fishing industry and the policy makers

The fishing industry, or more accurately, the catching sector is extraordinarily diverse (and herein lies the key to the policy dilemma). It is made up of a wide range of vessel sizes and *métiers* (or fishing methods). Vessels range in size from 6 or 7 m 'day boats', crewed by one, two or three men and confined mainly to inshore waters, through 15–30 m vessels with larger crew numbers depending on the type of fishing and more likely to be at sea for several days fishing for mixed demersal species in offshore waters, to the giants of the fishery, the modern pelagic (i.e. mid-water) trawlers of up to 80 m in length fishing migrant shoals of herring, mackerel and certain other species partly for reduction to fish meals and oils. Paradoxically these large vessels costing several million pounds sterling to build are usually at sea for less than half a year; and their complement of crew is often not all that much larger than the smaller, multi-purpose offshore vessels, thanks to the sophisticated technology for locating, catching and handling the fish. Apart from the highly specialised pelagic trawlers, most fishing enterprises are designed to target a variety of species, often combining a sequence of seasonal fisheries, each requiring a different *métier*; it is this operational versatility that forms the basis of their viability and resilience in face of changing circumstances.

Fishers are a relatively rare breed and declining in number. In the UK today there are around 12,000 and they may account for a mere 0.01% of the workforce though for every fisher at sea there are three others working with fish on land. Until quite recently there were probably as many fish and chip outlets throughout Britain as there are fishers catching the fish today! Despite the diversity that characterises Europe's fishing industries, there is one common defining occupational feature: the dangers, discomforts and risks that all fishers face in pursuit of their trade irrespective of vessel size or type of fishing. Fishing is widely recognised as one of the world's most dangerous occupations: it takes place in one of the least forgiving environments given to sudden changes of mood. It is not unknown for fishing to continue in quite heavy seas with the boat pitching and rolling making the task of hauling the nets – hopefully laden with fish – both arduous and sometimes dangerous, while others are engaged in gutting, cleaning and boxing up the catch. On the fishing grounds the crew may find itself working 12 or 15 hours a day in appalling weather conditions. And except for most modern, larger vessels, living conditions on board are usually cramped and spartan. Although the physical challenge may be less on the smaller, open 'day boats' operating in sheltered, inshore waters but with less protection from the elements, the hauling of static gears (fixed nets, lines or pots) usually by hand can prove difficult in poor weather and the return to port made less easy by small engine capacities.

And all of this is endured for a sometimes meagre, usually uncertain level of remuneration – with the hope rather than expectation of a good catch and high earnings from a trip. While the crews of the larger offshore vessels are paid a basic wage plus a share of the net profit from a trip, on the smaller boats the net revenue (once operating costs are deducted) is divided among the crew on a share-based system. It is not altogether uncommon for the skipper owner to be paying himself the equivalent of the minimum wage for a particular trip. Not surprising, therefore, to find that skippers customarily regard fishing as a vocation rather than a job.

In essence, the catching sector is made up of a myriad of family owned and operated vessels and occasional partnerships comprising members from two or more separate families. Increasingly, however, skippers are having to look beyond the confines of the family and the local community to recruit crew members. Corporate ownership of fishing units remains quite rare in Europe since the demise of the distant water fleets of Germany, Belgium, France and most notably the UK. Among the small and medium-sized enterprises, the skipper owner, who has usually served a long apprenticeship on his father's vessel and honed his nautical skills and gained his local ecological knowledge on which the success of his enterprise is largely based, values his independence and self-reliance very highly. But he, too, must rely on the ability of his 'shore crew' – traditionally his wife and/or other non-seagoing members of the household – to ensure that supplies needed for the next trip are readily available, accounts paid and any emergencies attended to without delay as turn around times of not more than a day or so are common. Not all fishing enterprises operate full time; for a variety of reasons, and especially in the case of small-scale fisheries, the owner-operator may choose to run his fishing operation on a part-time, seasonal or even occasional basis.

One of the persistent problems facing the catching sector overall – and smaller vessels in particular – is the weakness of its collective organisation. This is, in part, a consequence of the multitude of widely dispersed, independent enterprises whose human resources may be stretched to the limits and, in part, to the huge diversity of scale, *métier* and of degree of reliance on fishing for household income. At the start of the period under review, fishing associations were mostly locality based and involved in negotiating with local authorities on local, port-based issues or finding local solutions to particular problems. National associations, capable of representing the interests of the fishing industry over national policy related issues, such as the National Federation of Fishermen's Organisations (NFFO) founded in 1977 and the Scottish Fishermen's Federation

(SFF) founded in 1973, were brought into being precisely to provide channels of communication between the industry and the increasingly influential bureaucracies at member state and EU levels at a time when fishing was becoming more politicised and contentious.

By contrast, the nature of the second protagonist is nowhere near as diverse and complex as that of the fishing industry: the pen portrait of the policy maker is therefore less nuanced and somewhat briefer. He or she is part of a bureaucracy (or civil service) that is often seen as 'faceless' or 'remote' by those in the fishing industry. Only a small fraction of the bureaucracy, now universally regarded as essential to the smooth running of the state, is directly concerned with policy making. Rather larger numbers are required for detailed implementation, research and data collection, and monitoring of the policy.

Entry into the policy making élite echelons of the civil service will normally be through one of two channels: progression through the ranks or via graduate selection and fast-tracking of high-performing individuals following a relatively brief apprenticeship that involves the shadowing of senior staff across a range of activities in order to gain some understanding of the policy process as a whole. Fishing is generally considered a technical policy area requiring the input of informed specialists. But traditions relating to civil service culture vary across Europe from a reliance on recruitment of specialised expertise in the fields of biological or environmental sciences, economics and law to the classic Whitehall model of recruiting the 'best brains' irrespective of discipline through a competitive selection process and a lengthy 'apprenticeship' gaining experience in a wide range of departmental settings that is intended to produce a well-grounded decision maker capable of fitting into almost any policy area. Only a small handful of senior civil servants engaged in the formulation and implementation of fisheries policy, however, will have any hands-on experience of the realities of fishing as an occupation or way of life – a clear hindrance to any meaningful direct dialogue between the policy maker and the industry that can only serve to enhance the image of a remote and uninformed bureaucracy in the eyes of most fishers.

The situation is also complicated by the fact that for the past 45 years the European fishing industry has been in thrall to at least two and possibly more separate bureaucracies: the European Commission in Brussels wholly responsible for science-based management of the resource (i.e. the fish stocks); and the member state responsible not only for the implementation of CFP regulations but also retaining control of key policy areas notably the quota management system. In some member states, where domestic policy areas are devolved to regional administrations, there is the possibility of a third tier. What is perhaps most remarkable in this multi-tiered governing system is the virtual unanimity of opinion within the bureaucracies around the general thrust of fisheries policy and the universality of output controls (catch quotas) as the chosen means of controlling fishing activity. Whether this consensus is built around a belief that quotas are, in fact, the most effective means for addressing resource stability or simply the easiest to calculate, implement and monitor remains a moot point.

Occupying centre stage: the fish

Central to the narrative and the focus of attention for both protagonists are the fish, elusive and largely uncharismatic in character but nonetheless playing a crucial role in the unfolding drama. The waters of the North Atlantic are home to a wide variety of commercial fin- and shellfish species that have been exploited over the centuries with increasing intensity but still provide the basis for some of the world's most productive fisheries. This is especially true of the North East Atlantic where the North Atlantic Drift (a.k.a. the Gulf Stream) carrying warmer surface water into the high latitudes, allows the mixing of warm surface waters with denser, colder bottom waters aided by strong circulatory currents that facilitate the recycling of nutrients. This, combined with the easterly drift of plankton that forms the basic food for most marine animal life forms in their early months and, finally, the extensive areas of shallow seas on the continental shelf, all combine to create the essential conditions for a highly diverse and prolific mix of species.

Despite this diversity, a mere seven or eight species make up the bulk of the catch – cod, haddock, saithe, whiting, redfish, plaice, herring and mackerel. Of these, the first six are classed as demersal or bottom-feeding species that occupy well-defined territories linking their specific spawning, juvenile and adult feeding grounds. The adult feeding grounds are commonly shared with other commercial species forming loosely structured mixed shoals targeted by bottom trawling or by longlining. The last two species, herring and mackerel, are pelagic, mainly to be found in the middle layers of the water column. Much smaller in size and highly mobile, they each occur in separate shoals – huge in numbers and tightly bunched in structure – on their long migratory journeys to spawning grounds where they form the focus of a seasonal fishery.

Even though it will barely feature in the central narrative, mainly because to date it has evaded the close attention of the CFP, mention must be made of the shellfish sector. The two principal groups of shellfish – crustacean (mainly crabs and lobsters) and molluscs (mussels and oysters) – are harvested in contrasting ways. Crab and lobster are caught in fairly primitive traps or 'pots', set in increasingly lengthy 'strings' more usually in inshore waters and taken up every second day or so and then relaid on roughly the same sites. By contrast, mussels and oysters are the product of mariculture systems, still reliant on the normal functioning of the inter-tidal environment but where the beds are cultivated and from time to time restocked to ensure a reliable harvest of high-quality shellfish. For other shellfish different modes of harvesting are used: for Nephrops (a.k.a. langoustines), now the highest earning UK shellfish, offshore trawling on grounds often overlapping cod and haddock feeding grounds has become the norm, while scallops are occasionally still hand harvested by divers but more commonly subject to the controversial dredging of the natural scallop beds.

Though the quantity of shellfish landings is very much smaller than for demersal fish, the landing values for the shellfish sector have increased significantly in recent decades and may sometimes exceed those for the demersal or pelagic sectors, largely in response to the growth of strong middle-class restaurant culture throughout much of continental Europe.

Only two (possibly three) of the fifty or so important species occurring in the North Atlantic have earned the distinction of acquiring cultural identities: the iconic cod (*Gadus morhua*), the herring (*Clupea harengus*) and possibly the salmon (*Salmo salar*). Both cod and herring have long commercial histories dating back to the early medieval period; both have provided a significant source of income and a means of paying taxes levied by the church and state in the days of the Hanseatic League; and both have made valuable contributions to domestic consumption and foreign trade. Of the two, the cod has earned perhaps the greater, universal reputation as the bedrock of the Atlantic fishing industry and as the consistently most popular choice for human consumption. It has also formed the basis for possibly the very earliest commercial form of 'fast food'. Battered and deep fried along with chipped potatoes, it was retailed through a network of independent 'fish and chip' shops to be found in almost every working-class urban neighbourhood in mid-20th-century Britain.

The herring, known as the 'silver darlings' in Scotland, was a less reliable source of livelihood and income on account of its seasonality and a tendency throughout its history for strong and quite lengthy fluctuations in availability. During the boom years its impact on the lives of the fisher populations could be quite extreme: the pressure on available human resources at the peak of the herring season would be felt by all family members, not only those at sea but also those forming the shore crews. In Scotland in the late 19th and early 20th centuries, the fleet of drifters tended to follow the migrating shoals along the east coast of Britain, together with their families (wives and daughters), moving down the coast from port to port as a mobile workforce for undertaking the preparation and packing of the catches into barrels. The herring also made a greater impact on the traditional folk culture celebrated in stories, poems and song.

Whereas the technologies surrounding demersal fishing for cod (and kindred species) remained faithful to the basic model of bottom trawling but developed very significantly in terms of scale, vessel design and levels of sophistication (see Chapter 2), the pelagic sector underwent a remarkably radical transformation in the type of fishing enterprise involved, during the later part of the 20th century. Originally involving large numbers of smaller vessels that fished close to the coast as the herring moved towards their spawning grounds using seine nets to enclose a part of the shoal before hauling the bulging nets on board by hand, the development of power-assisted hauling gear in the 1960s began a radical transformation. Over a period of roughly 25 years, it resulted in the replacement of the vast fleets of small multi-purpose boats by a very much smaller fleet of less than a hundred giant pelagic trawlers operating from ports throughout western Europe: a very far call from the folk song image of herring fishing depicted in Ewan MacColl's *The Shoals of Herring* written as late as 1960 when the herring fishery was in its heyday.

An immediate consequence of the ability to haul much larger volumes of herring on board the growing fleet of larger boats was a surge in landings during the boom years of the 1960s and the inevitable diversion of parts of the catch to reduction plants for fishmeals and oils partly as a result of higher rates of wastage due to the herring's relatively rapid deterioration in quality after capture and partly in response to increasing demand for protein rich livestock feeds from the emerging intensive livestock systems. The status of herring as a food fish was being undermined. Rarely regarded as a staple source of food, unlike cod, herring was valued more for its contribution to a healthy diet as a source of unsaturated fats and relished as a speciality food in its various guises. Processing the fish was essential: it entered domestic and overseas markets in a range of traditional forms – dried, smoked, salted or cured.

There is a third candidate for iconic status. The Atlantic salmon has a very different lifestyle: described as an anadromous species it spends most of its life at sea, mainly in the north west Atlantic from where it will migrate to the upper reaches of relatively fast-flowing rivers in northern Europe (Ireland, Scotland, western Norway and Sweden) to spawn. The brood spends only a few weeks on the spawning ground before returning to the Atlantic where it will remain as juveniles for two or three years before joining the spawning adults in their great migration. The fame of the salmon owes just about everything to its role as a sports fish – the target for a traditionally wealthy elite of fly fishers – and as such may be considered a quasi-commercial fishery earning incomes for the riparian landowners, through whose lands the salmon rivers flow, by means of the sale of seasonal fishing licences rather than from the value of the catch. Salmon management has centred on the sustainability of stocks for the benefit of the sports fishing clientele. Offshore fishing for Atlantic salmon is now virtually banned and the number of licences granted to commercial fishing interests for coastal netting of the migrating fish in the UK has been greatly curtailed in recent years. Today, the main source of salmon entering the seafood supply chain is as farmed fish, the product of aquaculture rather than fishing.

In order to understand the reasons why fish were to prove such a difficult and at times volatile resource for policy makers to handle during the late 20th century, it is necessary to take on board some basic lessons as to their status within the marine food chain, their reproductive powers, their life cycle behaviour and the impact of fishing mortality. The first lesson is the simplest. As regards the food chain the basic rule is that 'big fish eat little fish ...', if not quite literally at least in the sense that during their life cycles – from egg, through the larval stage and into the juvenile stage – most species will be predated on by larger, more mature fish.

Second, most but not all species have prodigious reproductive potential with mature females laying up to several thousand eggs each year; fish also continue to grow in size and retain their fecundity throughout their adult lives. The reason for this very high reproductive potential is that fish also suffer very high natural mortality rates especially in their early life cycle stages. The timing of certain events can be crucial: for example, the emergence from egg into larva is timed to coincide and converge with the so-called 'plankton bloom' with eggs, larvae and plankton all being carried along by surface currents. Any failure in timing is likely to have a negative effect on the strength of the affected year class. Not all young-age mortalities are attributable to predation as large numbers of larvae will simply die from starvation.

As a natural resource fish populations are totally dependent on the normal functioning of a natural environment subject to minor variations that can cause short-term fluctuations in food supplies and be reflected in annual fluctuations in the strength of particular year classes that eventually determine the overall size and structure of the spawning stock. Possibly the most crucial life cycle change is the recruitment of juvenile populations to the adult breeding stock for it is this that ultimately determines the long-term sustainability of the fishery. Even in an unexploited state fish stocks will normally show a tendency towards minor annual variations and occasional longer-lasting and deeper troughs, as in the case of herring in the north east Atlantic in the second half of the 1900s. However, the crucial factor in determining the long-term sustainability of a population is fishing mortality which measures the impact of fishing activity on the stock. Under light or even moderate levels of fishing effort most stocks will retain their innate capacities for renewal, but if fishing effort continues to expand it will eventually reach a point where significant structural changes will be triggered. More of the larger, more fecund fish will be taken than can be replaced by recruitment from juvenile stocks and a state of 'recruitment overfishing' will occur; the situation will intensify as more of the smaller fish will need to be caught to maintain the overall volume of the catch.

Returning to the first lesson, the overall structure of the food chain will alter as a consequence of overfishing. The removal of large numbers of top predators like cod and haddock – the favoured targets of Atlantic fishing activity – will allow some prey species to thrive, expand their populations and fill the niches vacated by the disappearing predators. In some instances, fishers will be forced to turn their attention to these prey species, usually of less commercial value, in a process known as 'fishing down the food chain'. Among the species to benefit from reduced predation, however, are the high-value crustacean shellfish – crab, lobster and Nephrops – proving once again that where there are losers, you are likely to find some winners.

Finally, it is important to understand that over the course of the last 50 years the status of fish as a resource has been greatly altered. Originally it was considered a common property resource or *res communis* that, together with the long-held principle of the freedom of the high seas, permitted fishers to choose when, where and what to catch. In the late 1970s, following the establishment of 200 nm EEZs, fish became 'nationalised' – still a common property resource but held in trust by the coastal state responsible for both allocating fishing rights and regulating the fishery. Later, with the introduction of quota management systems that allowed fishers to buy, sell or lease their shares in the fishery, the resource became effectively 'privatised'.

Supporting roles: fisheries science and politicians

Two further actors, both in supporting roles, deserve special mention: fisheries science, playing a fundamental and substantive part in the policy process, and the politician, a surprisingly shadowy, less immediately influential character. Fisheries science is an established and respected branch of the environmental sciences with its roots in marine biology. Its role in fisheries governance is to furnish the scientific basis for policy relating to the management of the natural resource through advice based on continuous assessment of the state of commercial fish stocks. The evidence is derived from annual sample surveys and is ultimately fed through to EU and member state administrations. The strength of that advice comes not so much from the results of the national stock surveys *per se* but rather from the results of a mediation process conducted by the International Council for the Exploration of the Seas (ICES) established in Copenhagen in 1902, involving senior fisheries scientists throughout the North Atlantic region.

As an environmental science, fisheries science is something of a poor relation. Rather less is known about how oceanic habitats and species behave and interact, compared to terrestrial ecosystems. Our knowledge base of the hydrography, chemistry, biology and ecology of the oceans that govern the behaviour and performance of the world's fisheries is much less developed than the knowledge base for agricultural production and seems likely to remain so. The complement of fisheries scientists employed throughout Europe's universities and state-funded laboratories will continue to be very much smaller than the agrarian scientists working to support agricultural food production for one very good reason: while farming, especially in the more developed countries, is now a highly sophisticated techno-scientific activity increasingly reliant on the artificial stimulation of crop and livestock yields, capture fisheries remain essentially a hunting economy, completely reliant on the normal functioning of the marine environment. Marine aquaculture in Europe, while increasing its output, remains confined to a small range of species.

Politicians, whom we usually think of as the puppet masters, pulling the strings of policy making, play a surprisingly subdued role in the formulation and implementation of fisheries policy to the point where it is tempting to think of fisheries as an essentially apolitical policy issue. There are two possible reasons for this: first, the highly technical nature of fisheries policy is best left to the experts; and second, equally plausible, an explanation that recognises that fisheries make only a very small contribution to the national economy, accounting for around 0.3% of GDP within the EU with only a slight variation among member states, and providing very low and declining levels of employment. In terms of national politics, fisheries appear irrelevant, commanding little attention in election manifestos, occupying a minimal share of parliamentary business and normally attracting little attention from the media. Things may of course be very different in the arena of local constituency politics in some coastal regions. Nor are fisheries-related matters usually the stuff of party politics; it would be difficult to draw a clear distinction between the views of the different political parties over fishing and equally hard to discern any significant changes in domestic fisheries policy as a result of a change in the elected government. Where politicians do come into their own is in the defence of national fishing interests, most crudely revealed in the annual December meeting of the Council of Ministers to agree the detailed distribution of fishing opportunities between member states for the following year.

On the other hand, the fishing industries themselves have become more 'politicised' in recent years. As a largely self-employed workforce, lacking the support of a strong trade union, they have sought to draw attention to their problems through active demonstrations varying in their degree of militancy from relatively unintrusive 'parades' by flotillas of small fishing boats, scarcely distinguishable from the annual regattas, the more noisome dumping of rotten fish on the doorsteps of government buildings, to the disruptive, often prolonged and sometimes threatening blockades of the Channel ports by the well-organised French industry. While the major national federations have striven to maintain a strictly neutral position, some smaller associations have aligned themselves with more radical political groupings, as in the case of Fishing for Leave and its alignment with the right-wing United Kingdom Independence Party (UKIP) during the Brexit campaign.

The role of pantomime villain: fisheries policy

If there is one feature in the story that is about to unfold that is held responsible for all the ills that befall the fisher and the fishing industry, it has to be the fisheries policy itself, especially in the guise of the Common Fisheries Policy. Virtually all actors – scientists, fishers, politicians and academic writers – are quick to blame fisheries policy for falling fish stocks, a declining industry, reduced employment, narrow profit margins, shortage of crew …, all with good reason and plenty of circumstantial evidence to support the charge. Fisheries policy is very much the pantomime villain. There is just one flaw in their argument: the state of the fish stocks was already nudging towards crisis before the modern era of fisheries policy began.

As we shall come to understand as the drama unfolds, the scope of fisheries policy is rather too narrowly defined. It tends to regard fisheries as simply a problem of resource management to be tackled through the agency of regulating the behaviour of the fishers. The theoretical framework for fisheries management is constructed on the basis of defining the undefinable 'maximum sustainable yield' and the fictitious concept of 'economic man'. Yet those responsible for defining the policy have little understanding of the behavioural norms, values and aspirations of the fishers who struggle on a daily basis with the risks and uncertainties of 'living the fishing', 70% of whom are operators of small-scale, family-based enterprises.

A sub-plot: the environmental impacts of fishing

In the latter stages of the narrative a sub-plot emerges in which the fishing industry is subject to critical scrutiny from another potentially influential source and cast in the role of environmental vandal. The focus of this subject is the level of damage caused by past and present fishing activity; and it involves a new cast of actors: the environmentalist, public opinion and the media. Environmentalism is the activist, occasionally militant offspring of the environmental sciences as represented by international organisations such as Greenpeace and Friends of the Earth and reflected in numerous smaller groups. The position adopted by the environmentalist is usually well grounded in the relevant science but sometimes selective in its choice of evidence and quite often less well informed about the behaviour and activities of its target. Its actions are frequently eye-catching and intended to arouse and manipulate public opinion in support of its cause. Public opinion is vulnerable and quite often readily open to persuasion. Among the public at large knowledge and understanding of both the marine environment and the fishing industry and its contribution to national and global food security are at best modest and in most cases negligible providing a potentially fertile soil in which to cultivate opinion. But, once engaged, public opinion can prove a powerful weapon in persuading politicians to act.

Linking the two players – the environmentalist and public opinion – is the third actor: the media comprising television, radio and the press. During the second half of the 20th century the functions, style and influence of the media have undergone a remarkable transformation. In the 1950s, newspapers, as the name suggests, were considered the principal channel for communicating news, current affairs and information to the public on a daily basis. But this role has been surrendered to television, available round the clock and able to bring up-to-the-minute reporting from around the world, leaving newspapers to interpret, evaluate and comment on yesterday's news in the hope of helping to shape public opinion. Today the relatively rare incursion of the press into the fishing industry's role in the degradation of the marine environment in the search for an item of sensational news that may have escaped other media outlets has on occasion yielded unreliable or 'fake' news as a result of inadequate investigation.

Television, on the other hand, has helped to educate the public through well-crafted documentary programmes. In recent years the public has benefited from several lavish, widely franchised TV series on the state of our global natural environment and its wildlife in which fishing, alongside climate change and pollution, are identified as principal threats to the survival of habitats, ecosystems and individual species. For fishing it is an uneven contest: the weight of the scientific argument supports the charge with widespread evidence of damage to habitats and wildlife caused in part by fishing. The industry has to carry the burden of a previous history of misguided action often out of ignorance or necessity but also on occasion as a deliberate evasion of the rules. The substance of the sub-plot is likely to become more pressing in future (see Chapter 10).

A warning

What is often overlooked when recounting the events of the 1970s and 1980s is that most of the actors involved were entering a wholly new experience and their actions therefore must be judged accordingly. Not only was the legal framework – the Law of the Sea – profoundly altered with the introduction of EEZs and the responsibilities of coastal state governments totally reconfigured, but the concept of fisheries management was in its infancy. Essential elements of policy making – databases, capacities for monitoring and enforcement – were at best poorly developed. There was no readily available toolbox of appropriate measures, and little by way of previous experience to guide the policy maker; almost all decisions had to be made without the benefit of precedent. And for the fishers who hitherto had been a law unto themselves, enjoying a strong measure of freedom and independence, the rules of the game were radically being altered, hedging them around with restrictive regulations. The stage was being set in the 1970s and 1980s for a drama that is hard to categorise: part historical, part tragedy, part comedy and at times the theatre of the absurd but always experimental. That drama is played out in three acts against a backdrop that narrows from the extensive canvas of the northern north Atlantic, through to the continental shelf of western Europe, ending up with the much more confined area of the seas around Britain and Northern Ireland.

PART I
Seismic change

2

LOSS OF THE DISTANT WATER FISHING GROUNDS

Introduction

Part I examines the two sources of climactic change – the loss of distant water grounds and the development of the Common Fisheries Policy – that completely altered the political geography and management of fisheries in the North East Atlantic in the last quarter of the 20th century. These two entirely separate but overlapping events share a common theme – how best to manage the living resources of the sea in order to ensure their sustainability and provide a measure of equity and responsibility among those who lay claim to a share of the riches they bestow upon society. The timing of the events is crucially important to an understanding of what was to follow in terms of the options taken in relation to future fisheries management. Had the sequence of events been different, it is possible that a significantly different course of political decisions would have ensued.

We begin our detailed narrative in the 1950s and 1960s in the peak years of fisheries production and the heyday of distant water fishing in the North Atlantic in order to understand both the context and processes that were to lead, some 15–20 years later, to the first of the seismic shocks that would reshape the future history of fisheries in the North East Atlantic.

The early history of deep water fishing

The global fisheries resource is not evenly distributed around the world. As a rule of thumb, the northern oceans are more richly endowed than the southern oceans and shallow coastal waters yield a greater diversity and volume of catch than the deeper offshore waters. The North Atlantic is certainly one of the richest and most heavily exploited macro-regions, with a broad swathe of waters stretching in a north easterly direction from the George's Bank and Grand Banks situated between 42 and

DOI: 10.4324/9781003362913-3

52°N, beyond Iceland to the west coast of Norway (60–72°N) offering perhaps the richest and most reliable fishing grounds of all, where cod is king.

According to Kurlansky, the first Europeans to be tempted by the riches of distant water fisheries in the North Atlantic were the Basques from the southern part of Europe's Atlantic coast who as early as the 10th or 11th centuries were fishing for cod off the Newfoundland coast. There were good reasons for this: the Basques were maritime peoples with excellent if primitive navigational skills; they came from an area of the European coast not especially well favoured by fisheries; they had access to abundant supplies of salt necessary for the preservation of the cod; and salted cod was a popular feature of the regional diet. By the 16th century, the Basques had been joined by the Portuguese, the Bretons and a small number of ships from south-west England that overcame their prejudice against salted cod by disposing of their catches in mainland European markets.

Each group established a seasonal summer fishery lasting from April to September. In certain instances, some fishing crews chose to settle permanently in Nova Scotia or Newfoundland, leaving more space on the return voyage for the highly valued cargo. The method of fishing was simple: each day on the grounds, a flotilla of small, one- or two-man, 20-foot open 'dories' would be put over the side of the mother ship to jig for cod for up to 12 hours before being summoned back aboard by whistle. Despite the fact it was a 'summer fishery', the weather on the Grand Banks could turn very quickly from calm to squally conditions. But the greater danger came from the risk of thick fog that could descend without warning and last for several days in calm weather. Visual contact with the mother ship became impossible and the chances of being quite literally 'lost at sea' were high.

Surprisingly little interest was being shown in the distant water fisheries of the North East Atlantic off Iceland and northern Norway, though there had been some fishing off Iceland by both French and Dutch boats in the 18th century. Given Britain's great interest in the search for the North West Passage and its participation in the Arctic whaling industry, its very subdued interest in distant water fishing opportunities is all the more difficult to explain. One rather unconvincing argument was the unpopularity of salted cod among British consumers; yet salted or dried cod was being exported to west African and Caribbean countries as part of the infamous slave trade.

The age of steam

The breakthrough came in the 1880s and 1890s as a result of the much-delayed adoption of steam power in the fishing industry. Replacing sail with steam not only shortened the travel times between the home port and the fishing grounds bringing the distant water grounds of the North East Atlantic within easier reach of the major west European fishing ports. But it also allowed them to be exploited on a year-round basis. Moreover, there was a reasonable chance that most, if not all, of the catches in the seas off Iceland could, if well iced and carefully stored aboard, reach European markets in 'fresh' condition – given a certain degree of latitude in the interpretation of the term. Possibly the most significant fact about the introduction of steam power, however, was that it provided the catalyst for a major innovation in gear technology: the introduction of the otter trawl. Hitherto, offshore fishing activity had been virtually confined to line fishing or to the use of fixed nets. The beam trawl, where the mouth of the net is determined by the length of the solid wooden beam that forms its base, was unsuited to the deeper, less predictable, waters of the North Atlantic. With the means of guaranteeing constant, steady forward propulsion, not possible under sail but readily available with steam-powered fishing vessels, trawling became the more reliable and productive *métier*, especially with 'doors' of the new otter trawl positioned some distance away from the mouth of the net allowing for much greater volume in the size of the net. The increase in fishing capacity was immediate.

The impact on the UK fishing industry, and especially on the relatively newly established centres of fishing activity for the exploitation of the rich North Sea grounds around the Dogger Bank, were not long delayed. Between 1885 and 1895 practically all 90 sailing smacks based in the Humber ports of Hull and Grimsby had been replaced by steam-powered trawlers. Soon, the slightly larger Hull vessels (around 30 metres in length) had been redeployed to Icelandic waters, reducing pressure on the North Sea grounds which retained the interests of most Grimsby vessels for a while longer. By the end of the 19th century practically all of the Atlantic coast fishing nations of western Europe were engaged in distant water fishing activity.

Although, at the time, the overall level of fishing effort being applied on the distant water fishing grounds was scarcely enough to make any noticeable impact on fish stocks, there was growing concern within the fisheries science community as to its likely long-term effects. While Holt in 1895 argued that fish stocks must be given the chance to spawn at least once before being exposed to the risk of capture in order to maintain the innate capacity for the renewal of the stock, Petersen in 1894 suggested that even in heavily fished populations, the reasons for the eventual decline in fishing activity had more to do with the numbers of smaller fish that would need to be caught in future to replace the larger fish that had already been removed from the stock. Such concerns were publicly eclipsed by the greatly revered Victorian scientist T. H. Huxley's soothing words at the time of the International Fisheries Exhibition held in London in 1883 to the effect that: "Any tendency to over-fishing will meet with its natural check in the diminution of the supply, ... this check will always come into operation long before anything like permanent exhaustion has occurred". While Petersen's 'growth theory' was to influence both fisheries science and economics in the 20th century, Huxley's message was to prevail on both sides of the Atlantic in the policy debate well into the second half of the 20th century.

The first half of the 20th century was a period of consolidation rather than transformational change, punctuated by the two world wars and hampered by the depressed state of the economy in the 1920s and 1930s. During both world wars distant water fishing was disrupted, initially by the secondment of the newest British trawlers to the Royal Navy for use as minesweepers, and in the second by the appropriation of much of the North Atlantic as an active war zone. Recovery after World War I was slowed almost to a halt by the recession of the 1920s. In Britain, Aberdeen – the centre of the UK herring industry and with interests in the distant water demersal sector – suffered the worst; Grimsby, one of the leading demersal ports dividing its focus of activity between the North Sea and Icelandic waters, fared somewhat better.

Only Hull appeared to make a qualified success of the opportunities available in the inter-war years. Its stronger company structure, with three quarters of Hull trawlers in the hands of companies owning more than ten vessels, made it more competitive yet still aware of the benefits of cooperation over the building of a shared ice-making and reduction plant. The modernisation of its fleet was an ongoing concern: Hull trawlers were, in general, newer, faster and larger (at around 40 m), with crews of 14 men, and were being deployed over a wider geographical area of the north east Atlantic including both the Barents Sea and, venturing deeper into Arctic waters, Bear Island (31°N). But the design of the Hull trawler remained faithful to the model of the 1890s: the traditional steam-powered, sidewinder trawler. Apart from the introduction of the diesel engine in place of steam in the 1920s, which was eschewed by the Hull owners, innovation was confined largely to the shore-based processing industry with the introduction of both filleting and freezing equipment also in the 1920s.

A measure of the consolidation around distant water fishing activity that occurred in Britain in the first half of the 20th century is the rise in the share of the overall UK catch taken from distant waters from an eighth in 1913 to a third by 1938. It was not, however, a particularly profitable period for the companies that owned the fleet: quayside prices were depressed by poor market demand, aggravated by oversupply; take-home pay for crew members, still largely dependent on 'poundage', fell; and low or non-existent profit margins led the owners to cut both basic rates of pay and crew numbers in 1931.

The 1950s and 1960s: the beginning of the end

Growth and innovation

The recovery of Europe's deep water fishing industry in the aftermath of World War II was quite unlike that in the 1920s and 1930s. It was immediate, unimpeded, continuous, bold, innovative and, as it proved, over the top. The 1950s and 1960s were without doubt the peak years of distant water fishing activity in terms of the number of participating countries, the size of the trawler fleets involved, their combined fishing effort, catch levels and revenue. But pressure on the finite, if renewable, resource was building towards breaking point. When it came, the end was quick, comprehensive and disastrous for nearly all fishing interests.

Initially, the recovery was focused on replacing the remnants of the pre-war trawler fleet with an improved and lengthened version powered by the cleaner diesel engine. Once again, however, the conservative (and cost-conscious) mentality of the Hull trawler owners prevailed and steam remained a preferred source of power; Britain, of course, still had access to seemingly vast resources of coal. Warner, in his fascinating account of fishing in the North Atlantic (*Distant Water:*

The Fate of the North Atlantic Fisherman, 1983) based on trips made on a range of different types of distant water vessels in the 1970s, described the 1950s as the golden era for the traditional wetfish (or 'fresher') trawler. Catches were increasing year on year throughout the region from the Grand Banks in the west to the Barents Sea in the east (Figure 2.1). The distance, and more especially the time taken, to return the vessel and its catch to the quayside market was becoming more of a problem. The combination of around ten days fishing and six days return trip from the Barents Sea meant that the early part of the catch was around 16 days old by the time the vessel berthed. On landing the auctioneer graded the catch according to freshness. Even with the best onboard storage much of the catch was subject to spoilage and in some instances poor onboard conditions could mean as much as 75% of the catch being condemned and sold at much lower prices for fishmeal.

The solution – in the form of the freezer trawler – was pioneered in the late 1940s and early 1950s by Salvesen's, a Scottish-based firm with interests in whaling. According to Warner, Salvesen's version was a vessel of around 2,600 GRT, 80 m in length, with accommodation for some 89 crew (fishers and processors) and with the ability to stay at sea for around 70 days. These specifications completely dwarfed anything currently involved in the fishing industry. In design it combined three distinctive innovatory practices: stern loading in place of the more dangerous side hauling of nets; automatic filleting machines instead of hand filleting; and the capacity to freeze the entire catch to preserve the quality of the product.

Fairtry I – a factory ship in the fullest meaning of the term – was launched in Aberdeen in 1952 and quite soon Salvesen's had three such vessels working the North Atlantic grounds. Very highly specialised in design, their profit margins depended on the availability of large quantities of cod (or similar gadoid species) of a size and shape suitable for the filleting machines. Their success therefore depended on standardisation, reliability of the catches and stability of consumer markets in terms of preference (for frozen cod), steady levels of demand and buoyant prices.

Despite their proven success, there were few takers for the factory ship scale of operation. The exception was the Soviet Union, a relative newcomer to the distant water club. They had two factory ships, the Pushkin and Sverdlovsk – both remarkably similar in design to the Fairtry model – fishing off Greenland and the Grand Banks; a few years later there were 35 Soviet factory trawlers operating in the north west Atlantic. Few other fishing nations opted for the factory ship model, preferring a smaller version – the freezer trawler – closer in size to the traditional fresher trawler but still combining two of the transformative design structures – stern loading and freezing – with a complement of between 28 and 32 crew members. Among the earliest investors in the freezer trawler was West Germany.

In 1961 the experimental *Lord Nelson*, a large stern trawler (c. 70 m), joined the Hull fleet. Hybrid in design, it was equipped with a fishmeal plant and freezing capacity for part of the catch but landed primarily wetfish. It was not until the mid-1960s that the 'thoroughbred' freezer trawler started to play a major role in the Hull fishing industry. Once again, the Hull owners had been slow to respond to a major innovation, only this time they almost left it too late.

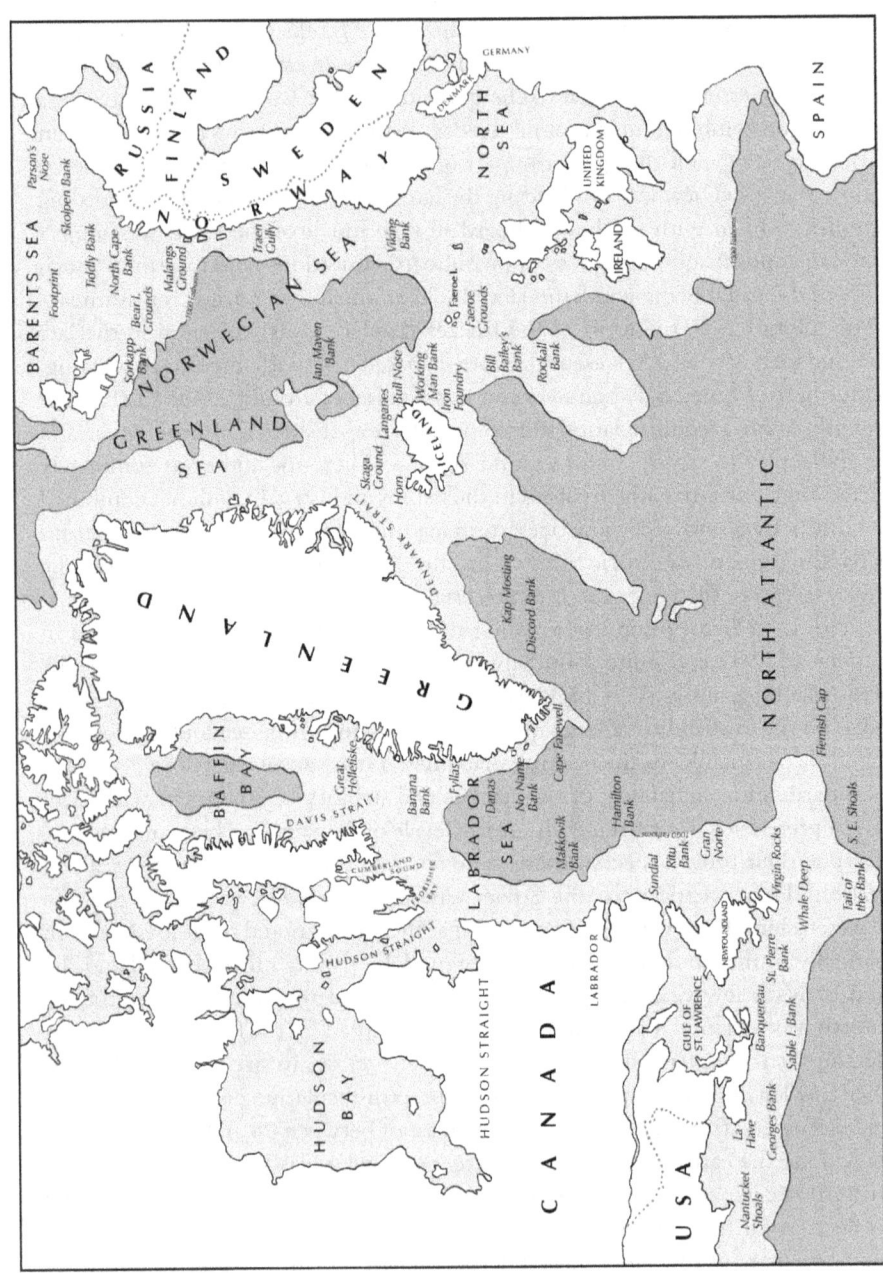

FIGURE 2.1 Northern waters (based on Warner, 1983)

Distant water trawling: an extreme occupation

A quite remarkable, stark and highly critical account of the Hull distant water trawling industry is provided by Jeremy Tunstall's *The Fishermen: The Sociology of an Extreme Occupation* (1962). In his introduction, Tunstall portrays the industry as offering 'a casual occupation … with a very high death rate, low pay per hour, appalling working conditions … and … forced retirement at an early age' (p. 11). He goes on to describe how, over the course of a year, the Hull fisherman 'spends perhaps 130 days and nights … working eighteen hours a day, another 120 days steaming at sea working eight hours a day. Ashore he has perhaps 35 days of "active" leisure between trips, and another 80 days or so of "trips off"' (p. 12), that is in temporary unemployment. The 18–20 man crew (skipper, mate, four engineers, radio operator, two galley staff, ten deckhands and a 'deckie learner' or apprentice) were nearly all recruited locally and the turnover among 14–15 year old 'deckie learners' was unsurprisingly high. The skipper had worked his way up from 'deckie learner' to mate and beyond before taking his master's ticket. Take-home pay for all crew members was based very largely on fixed shares in the net profit for the trip: it, therefore, varied greatly between the best and the worst performing vessels with the top skippers, often in command of the latest and best-equipped trawlers, able to select the more reliable crews.

Fishing throughout the year – and around the clock – distant water trawling was perhaps the most physically strenuous and dangerous of all occupations. Risk of serious injury, heightened by sheer fatigue, or of being swept overboard by a rogue wave, was constant. Fishing in the high latitudes in the winter months exposed the trawler to the greatest danger from severe storms, huge waves and above all 'icing up' when the superstructure, decks and rigging quickly became coated with thick layers of ice putting the stability of the ship at risk. Incidents involving the loss of the trawler with all hands were not uncommon.

Three events, each involving Hull trawlers, encapsulate the dangers implicit in fishing the waters of the northern Atlantic in storm wracked winter months. The first was the sinking of the *Roderigo* and *Lorella* off the north west coast of Iceland with the loss of all 40 hands on 26th January 1955 in 'hurricane force winds' after both vessels had iced up; the second, again off north west Iceland, occurred in 1966 over the space of three weeks when three trawlers – the *St. Romanus* on 11th January, *Kingston Peridot* on 26th January (both from Hull) and *Ross Cleveland* (4th February) from Grimsby – were overwhelmed by severe and prolonged storms with the loss of 58 lives and only one survivor. Circumstances surrounding the loss of the *Gaul* with all 36 hands in February 1974 were altogether different. Whereas the previous events had each involved traditional sidewinder trawlers, the *Gaul* was one of four newly commissioned stern hauling, freezer trawlers regarded as among the most modern, well equipped and safe vessels in the British fleet. Its sinking off Norway's North Cape in severe weather following 'a sudden loss of radio contact' gave rise to considerable media speculation but has never been fully explained.

In addition to the insecurity of employment, the risk to life and limb, and the wear and tear on the trawlerman's mental and physical wellbeing, there was no pension provision; it was, after all, formally regarded as a casual, seasonal form of employment. Early retirement, in their 1940s and 1950s, was commonplace. Some were able to find less arduous jobs at sea; others moved into shore jobs related to fishing or into unrelated manual labour. But many ended their working lives on the dole – a poor recompense for the years spent in one of the most physically demanding and dangerous of occupations.

The tragedy of the commons

Disquiet among fisheries scientists over the impact of the latest onslaught on the resources of the North Atlantic's distant water fishing grounds was running high but there was no discernible easy way out. International organisations like the International Commission for the Northwest Atlantic Fisheries (ICNAF) and the North East Atlantic Fisheries Commission (NEAFC), established in 1949 and 1959 respectively with the aim of finding some means of asserting a degree of control over rising pressure on resources were powerless to intervene. Membership of the Commissions was voluntary and any attempt to regulate fishing in international waters was nullified by the fact that only the assenting member nations were obliged to carry through the agreed regulations. The high seas were virtually ungovernable. Although the combined efforts of academic fisheries scientists and economists had succeeded in creating a single bio-economic model in the late 1950s that perfectly described the interactions between fish stock response to increasing fishing effort leading inevitably to the depletion of the stock and declining profitability of fishing operations, no prescription was available as to how to avert the looming crisis.

The most telling contribution to the academic debate was Garrett Hardin's 'The tragedy of the commons' published in *Science* in 1968. He likened the oceans to the commonages of pre-industrial farming communities where the small tenant farmers, limited in their opportunities for generating cash incomes from their in-field land, tended to overstock the commons with livestock (cattle and sheep) and, despite the decline in the carrying capacity of the overgrazed commons, sought to increase output by adding yet more livestock. His analogy was flawed in its historical assumptions both in respect of the behaviour of pre-industrial communities and also in the assumption that open access is necessarily associated with common property regimes in fisheries. Access to the common pastures was usually subject to strictly observed rules governing the numbers of grazing livestock; and often it was the landowner who later took ownership of the commons and overstocked them with sheep or alternatively managed them for shooting rights. Moreover, many pre-industrial societies developed quite sophisticated sustainable management regimes for their inshore fisheries, but were powerless to prevent the incursion of foreign vessels fishing without let or hindrance. However, Hardin's central message was unchallenged: the root cause of overfishing was the inability of open access regimes in international waters to prevent the situation where 'too many boats were chasing too few fish'. Ultimately, for fishing effort to be regulated the commons had to be enclosed and responsibility for their management to be reassigned.

The early 1970s: enclosing the commons

According to Warner, by the mid-1970s there were in excess of a thousand West European and Soviet bloc vessels active in the distant water grounds of the North Atlantic accounting for a total catch of 2.2 m tons – more than the New England, Canadian east coast, Icelandic and Norwegian fishing industries combined – but a little below the peak landings of 2.4 m tons reached in 1968. The 'occupation' of these waters by foreign fishing interests was all but complete. However, catch per unit of effort was declining and the size of the fish that made up the catch was getting smaller, both classic signs of overexploitation. A crisis point had been reached. By the early 1970s, the sovereignty of the high seas was being challenged. In 1973 a session of the UN Seabed Committee was convened, attended by some 34 nations mainly from the global South including Latin American, African and Asian coastal states together with two northern European nations (Iceland and Norway). The committee endorsed, in principle, the concept of a 200-mile exclusion zone but at the time no further action was proposed.

It was therefore left to a small island nation, Iceland (population c. 230,000), literally at the centre of the crisis, to take the initiative. No country within the North Atlantic basin was so dependent on its fishery and no fishery was so clearly threatened by the concerted growth of international fishing effort. The development of Iceland's fishing industry in the post-war period had initially been slow and uncertain, led by the replacement of its own small trawler fleet by newer, larger vessels owned either by a local community cooperative, the municipality or by the local processing plant. By the early 1960s a fleet of around 30 conventional sidewinding trawlers was landing a substantial share of the Icelandic catch but it was already suffering from high operational costs due in part to the depletion of stocks and unusually high operating costs. A second, more significant phase of modernisation was the introduction in the early 1970s of the more efficient stern wetfish trawler and by the end of the decade most of the local fishing centres found mainly around the western and northern coasts of Iceland, each with their own freezing plant, were served by at least one stern trawler.

Post-war Icelandic governments had sought to protect their national fishing interests through unilateral action to extend its fishing limits: first in 1952 to 4 nautical miles, again in 1959 to 12 nm, in 1972 to 50 nm and finally in 1976 to 200 nm. Although all four actions flouted the norms governing the freedom of the high seas principle, neither of the first two had any material impact on distant water fishing. The extension from four to 12 miles, regarded by the British as 'unwarranted under international law', was the first to earn the soubriquet of a 'cod war'. While six European fishing nations (France, Belgium, the Netherlands, West Germany, Denmark and Spain) all initially formally supported the British in their opposition to the new limit, by 1958 all foreign trawlers apart from the British had withdrawn beyond the 12-mile limit. The British continued to fish within the 12-mile zone in rectangular 'boxes' under the nominal protection of Royal Navy vessels. It was not until 1961 that the UK finally acceded to the 12-mile exclusion.

The second and third 'cod wars', both prompted by evidence of declining stocks and the need for urgent conservation measures, were much more provocative and dangerous in the tactics deployed though remarkably no lives were lost. They were provocative in the sense that they very clearly breached all known protocols concerning the long-established freedom to exploit the fishery resources of the high seas in first declaring a 50 nm limit in 1972 and later in 1975 a 200 nm exclusion zone with very significant repercussions for distant water fishing interests. And they were potentially dangerous because of the 'battle-plans' drawn up by the Icelandic Coast Guard. Clearly outgunned by the superior firepower of the British navy vessels that accompanied the British trawlers, the tactics adopted by the Icelandic Coast Guard vessels proved highly effective. Having first warned the trawler skipper to leave the 50-mile zone and having received no confirmatory reply, the coast guard vessel would alter course to sail between the trawler and its net deploying a device that first entangled then cut the wires connecting the trawler and its trawl, causing the loss of both the net and its catch and leaving the trawler owner with no option but to abandon fishing. According to Kurlansky, within a 12-month period in 1972/1973 no fewer than 69 British and 15 German trawlers lost their nets as a result. The third, final and decisive 'cod war' was, in fact, the shortest. In a matter of months, during which the Icelandic Coast Guard had succeeded in cutting a further 46 British and nine German trawls, and prompted a period of intense negotiation, the resistance was over.

Some distant water fishing did continue after 1976. The following year the UK was granted a derisory quota and licence for a handful of vessels to fish off Iceland. Some of the newest recruits to the growing Hull fleet of freezer trawlers fished off North Norway in 1977, playing 'hide and seek' with the Norwegian coast guard before eventually being escorted out of the territorial waters. But the damage had been done; under the new restrictive conditions, it was proving difficult to make distant water fishing pay.

Reason, respect for the natural rights of small nations and a concern over the sustainability of the fisheries in one of the world's richest fishing regions had finally prevailed over an old-fashioned sense of colonialism, naval supremacy and an ill-informed view that 'the sea hath fish enough'. The archaic interpretation of the principle of the freedom of the high seas had been challenged and Iceland had unilaterally secured its own *de facto* 200 nm fishing zone. The achievement was to prove highly infectious and by the end of 1976 almost all coastal states within the North Atlantic had followed suit. Action now moved from the North Atlantic to the United Nations in New York.

UNCLOS III 1982

Translating a *fait accompli* into international law was a more long drawn-out affair, though in fairness there was far more than fishing rights at stake. There was little danger that the principle of a 200-mile exclusive fishing zone, effectively covering all fishing activity across the continental shelf, would be overturned but defining the rights, responsibilities and obligations of the coastal states in respect of that zone to the satisfaction of the great majority of UN member states was a delicate task. Moreover, the definition of rights etc in respect of other uses of the sea and seabed within such an exclusion zone – notably with reference to hydrocarbon energy resources (oil and gas), renewable energy sources and navigation rights *inter alia* – further complicated the issue. When UNCLOS III was finally introduced in 1982, there was still some doubt as to whether it would receive its ratification some ten years later.

It is important at this juncture to take a closer look at the conditions set out in UNCLOS III for these will have a bearing on the later stages of our narrative. UNCLOS III was published in December 1982, setting out protocols for the conduct of all aspects of maritime activity carried out within the 200-mile Exclusive Economic Zone (EEZ): the protocols relating to fisheries were contained in Articles 55–75 of the Convention. The overall effect is to create an entirely new set of rules concerning the rights, responsibilities and obligations of coastal states in respect of fisheries management within the EEZ. It does not, however, wipe the old slate completely clean.

While the coastal state is granted sovereign rights for 'exploring … exploiting, conserving and managing' the fisheries resource (Article 56), and is obliged to 'determine the allowable catch of the living resources … ensure [the] maintenance of those resources is not endangered by overexploitation' and where necessary take action to restore the status of the fish stocks to maximum sustainable yields (Article 61), it cannot automatically claim the allowable catch for its own exclusive use. In promoting the 'optimum utilisation of the resource', the coastal state is obliged first to determine its own capacity to harvest the resource and then 'afford other states access to any surplus allowable catch' taking into account the needs of other states that have historically fished the area of the EEZ (Article 62). Finally, where stocks are shared between the EEZs of two or more coastal states there is a requirement for those states to 'agree upon the measures necessary to coordinate and ensure the conservation and development of such stocks' (Articles 63).

Clearly, UNCLOS III came too late to define the basis for the settlement of access rights for foreign vessels following the unilateral declaration of sovereignty affecting the North Atlantic fisheries. Limited access rights were, in fact, granted by both Iceland and Norway though not to a level where the concept of distant water fishing activity remained viable. As we shall discover later in the narrative, however, the calculation of 'allowable catches', 'access rights' for foreign vessels and 'shared stocks' – all new concepts promoted by the enclosure of the commons – are potentially contentious issues.

The impact of losing the distant water grounds

It is no exaggeration to assert that enclosing the commons brought to an end a significant era in the development of modern fisheries – one that, as a result of its uncontrolled intensity, had begun to threaten the long-term sustainability of the resource and suffocated attempts by the coastal states to develop the full potential of their own domestic fishing industries. The resulting redistribution of costs and benefits was quite simple: the benefits in terms of reduced pressure on the fish stocks were enjoyed almost exclusively by the coastal states; the costs, in terms of lost fishing opportunities, were borne by the distant water fishing nations of both western and eastern Europe.

Not only did enclosing the commons compel a radical restructuring of the European fishing industry after 1975 and a reorientation of regional trade in fish and fish products throughout the North Atlantic basin, but it also created a new balance of power within the North Atlantic between the declining 'industrial' fishing nations like the UK, Germany and Belgium on the one hand, and new fishing powers of Iceland, Norway and to a lesser extent Greenland and the Færoes, on the other. But perhaps its greatest impact lay in creating, for the very first time, a genuine opportunity to actively manage the fisheries in ways that should ensure the sustainability and optimal use of the living resources of the sea as ecological, economic and socio-cultural assets. Sadly, as the unfolding narrative will reveal, some 45 years later in certain parts of the region the management approach remains sub-optimal.

The impacts arising from the loss of distant water fishing opportunities were felt at national, regional and local levels and in a variety of ways. One of the countries to suffer those impacts most keenly was the UK, though it can very easily be argued that proportionally the Federal Republic of Germany and Belgium were the worst hit, if only because the new political geography of fishing left them much less able to find compensation within their own very restricted EEZs. In the case of the UK, a high proportion of its active fishing capacity became virtually redundant and, with roughly 30% of its total 'domestic' supplies now cut off, what had once been a largely self-sufficient seafood economy now became heavily dependent on imported supplies of fish. Fortunately, the ending of distant water fishing overlapped the initial phases of the globalisation of trade in fish and fish products; indeed, the former may well have acted as the catalyst for the latter. To make matters worse the readjustment of the UK's fisheries sector to the loss of distant water fishing opportunities was hampered by the fact that any moves to reform future development on expanding fishing activity within the UK's own relatively extensive, well-endowed fishing zone were effectively blocked by the membership of the EEC and the basic principles governing the development of the CFP (see Chapter 3).

Despite the dominant influence that fishing held in the industrial images of both Hull and Grimsby, in terms of direct development fishing related jobs accounted for around 17% of the labour market in Grimsby and a mere 5% in the much larger city of Hull. The loss of seagoing and quayside jobs was nevertheless a severe shock to the local economy and cultural identities of both ports. However, the bulk of fishing-related jobs was protected by their roles as the principal wholesale wetfish markets controlling the inland distribution network and the continuing strength of the Humberside fish processing sector. Direct landings by local trawlers were superseded initially by landings from foreign trawlers but ultimately through regular consignments of containerised fresh cod and haddock from Iceland, less frequent deliveries of fresh and frozen fish from Norway, the Færoes and, occasionally, Russia, and small amounts of fish overlanded from other British ports.

The events of the mid-1970s did not imply a complete break with the distant waters of the North Atlantic. Indeed, in 2019 a new state-of-the-art factory trawler – the 81.2 m *Kirkella* became the latest addition to the much-depleted Hull distant water fleet. With highly automated equipment for gutting, filleting, grading, packaging, freezing and palletising its catch, it is capable of delivering 780 t of frozen fillets – the equivalent of 2.3 million 'fish suppers' – at the end of a typical six-week trip in the high latitudes of the North Atlantic.

What became of the beneficiaries?

It is possible that the actions taken around the middle of the 1970s came just soon enough to 'save' the iconic cod from the danger of collapse as a viable commercial fishery. For some time there had been clear indications that it was already being seriously overexploited and there was no way of knowing how much pressure it could sustain. The political framework, within which responsibility for ensuring the future welfare of fish stocks was defined, had been set up and it fell to a handful of newly empowered coastal states – Canada, Norway and Iceland together with Denmark's self-governing territories of the Færoes and Greenland – to take up the challenge of managing depleted fisheries in the high latitudes of the North Atlantic. Each of the main beneficiaries opted for quite different strategic approaches to the development of their domestic fishing industries and, by inference, to the sustainability of the resources in their care.

The five smaller countries all had very good reasons for wanting to see their domestic fishing industries expand and prosper. In terms of revenue, employment and external trade, fishing already played a significant role in the overall national economies and the potential for future economic growth appeared to lie in further utilisation of the living resources of the sea. Partly because of their small size and greater economic dependence on the fisheries much closer relations were forged between the governing institutions and the industry such that in Norway policy making was aptly described as a co-management system based on 'centrally directed consultation'.

Over the years Norway has earned a reputation for responsible management of its fisheries. Initially, it continued to support its coastal fisheries and the welfare of its dispersed coastal populations especially in the north of the country within a framework of sustainable resources rather than encourage a vigorous restructuring and greater concentration of its fishing industry. More recently, under pressure from the Nordic welfare model based on combining private prosperity with high taxation, Norway's policy has become rather more aligned to a neo-liberal market approach while avoiding some of the more extreme social costs that this might entail. Here, Norway's careful husbandry of the wealth generated from North Sea oil development may have helped. In the immediate aftermath of 1976 Norway successfully engineered a lasting agreement with the Soviet Union over the allocation and sustainable use of the shared fishery resources of the Barents Sea, involving close cooperation within the fields of research, regulation and compliance control.

In marked contrast, Iceland's initial response to the new opportunities in the 1970s and early 1980s was a rapid and somewhat indiscriminate investment in both the fishing fleet, involving new builds and as a stop gap measure the purchase and refitting of displaced distant water trawlers, and new or expanded processing plants. While this expansion in fishing capacity was never going to equal pre-1975 levels of foreign fishing effort, it made the recovery and conservation of its resource base less secure. Within a few years, however, Iceland's fisheries sector was to undergo a major transformation following a two-stage adoption of a market-based system of management. The first intervention involved the introduction in 1984 of individual transferable quota (ITQ) for the cod fishery, allowing quota holders to buy, sell or lease quota among themselves, and later in 1991 the extension of the ITQ system to all fisheries thereby establishing Iceland as only the second developed coastal state to embrace fully a market-led approach to management. This had three main effects: first, it led to the improved performance and efficiency of the harvesting and processing sectors through radical structural and spatial concentration; secondly, it effectively ended indiscriminate investment in increased fishing capacity, reducing pressure on the resource and allowing more effective resource management; thirdly, it had severe negative effects on the inshore sector and on fisheries based employment particularly in the many small settlements around the coast that lost both their freezer plants and trawlers in the restructuring process. In 2004 Iceland introduced a separate quota management system for the small boat sector, based on individual transferable quota, that according to Dobeson (2018) disentangled the sector from its ties to the local processing plant and led to the development of a more entrepreneurial approach in a more competitive market environment. The debate continues as to whether the overall economic gains outweigh the considerable social costs, but the medium-term viability of the Icelandic fisheries sector remains assured.

Of the six coastal states to benefit directly from enclosing the commons, it is the comparatively wealthy, developed and largest country with a potential to nurture and sustain the rich marine resources off its eastern seaboard that provides the enigma. Canada appears to have taken its eyes off the ball and plunged the coastal economies of the maritime provinces – especially Newfoundland – into deep depression through the mishandling of its fisheries resource. Precisely why this should happen is hard to explain.

Writing in the late 1980s, Sinclair presents a picture of conflicting fortunes for the prospering offshore fisheries undertaken by a modern trawler fleet owned mainly by shore-based processing firms, on the one hand, and the nearshore longlining fleet and the struggling inshore sector of small scale fishers reliant on summer season static gear fishing (cod traps, lobster pots) and hand jigging, on the other. This combined with non-fishing related employment locally or on the Canadian mainland was expected to provide an adequate income for the participating families. Small-scale vessel owners were also caught up in a persistent cost-price squeeze, involving increasing costs of production and depressed prices for the fish.

Underlying this situation there were questions raised over the stability of the overall catch levels and the competence of the government departments involved in setting and allocating the annual total allowable catches (TACs) for groundfish species, especially the all-important cod. In particular, there were concerns over the reliability of scientific stock assessments and the advice given to government over both short- and medium-term prospects. Certainly, official confidence in the state of cod stocks was to prove tragically misplaced. Owners pointed to over-reliance on allocating uncertain TACs as firm individual vessel quota, the incentives given to the fishers to catch more fish than their quota warranted and then adjust their final landings by discarding poor quality fish before landing, and the persistently adversarial relationships between government officials and the fishers. Whatever the cause, by the early 1990s cod stocks on the Grand Banks were in a very poor condition and in 1992 the Newfoundland cod fishery was closed. The lifeblood of the many small communities ('outports') around Newfoundland's coastline simply drained away. Recovery has been uncertain and long drawn out with only a small number of fishers able to find sanctuary in the now booming lobster fishery.

For the other North Atlantic states enclosing the commons has worked out, as was originally intended, with soundly based resource conservation policies complemented by the development of thriving and dynamic export-oriented fisheries sectors filling the role left vacant by the departure of foreign distant water fleets. In the next chapter, we turn to look at how western European fishing industries sought to cope with the changed political geography of fisheries through a transformational process that was as radical and traumatic as the loss of access to distant water fishing grounds but more nuanced and possibly more covert in the achievement of its goals.

Further reading

Dobeson, A. (2018). Economising the rural: how new markets and property rights transform rural economies. *Sociologia Ruralis*, 58 (4), 886–908.

Kurlansky, M. (1998). *Cod: A Biography of the Fish that Changed the World*. London: Jonathan Cape.

McGoodwin, J. R. (1991). *Crisis in the World's Fisheries: People, Problems and Policies*. Stanford: Stanford University Press.

Sinclair, P. R. (Ed.). (1988). *A Question of Survival: The Fisheries and Newfoundland Society*. St John's: Memorial University Press.

Symes, D. (Ed.). (1987). *Humberside in the Eighties: A Spatial View of the Economy*. Hull: University of Hull, Department of Geography.

Symes, D. (Ed.). (1997). Fisheries management in the North Atlantic; National and regional perspectives. *Special Issue of Ocean and Coastal Management*, 35 (2–3), 51–224.

Symes, D., & Phillipson, J. (2016). Industrialising the marine commons: adapting to change in Europe's coastal fisheries in Shucksmith, M. and Brown, D. L. (eds.) *Routledge International Handbook of Rural Studies*, pp. 323–334. London: Routledge.

Tunstall, J. (1962). *The Fishermen: The Sociology of an Extreme Occupation*. London: McGibbon and Kee.

United Nations. (1982). United Nations Convention on the Law of the Sea, available at <http://www.un.org/depts/les/convention_agreements/texts/unclos/UNCLOS-TOC.htm>

Warner, W. W. (1983). *Distant Water: The Fate of the North Atlantic Fisherman*. London: Penguin Books.

3

THE MAKING OF THE COMMON FISHERIES POLICY

A change of scenery

The setting now changes from the turbulent open waters of the North Atlantic, a backcloth for the toughest and most dangerous occupation imaginable and behaviour that some may regard as akin to piracy, to the more confined spaces that make up the domestic fishing grounds of western Europe and the shadowy corridors of power in Brussels. Here among rumours of conspiracy and double-dealing, the future of the fishing industries of a growing number of coastal states were about to be decided. Although the settings are very different, the time frames of the final years of distant water fishing and the acceptance of a wholly new political framework for fisheries are more or less the same. The second half of the 1960s and the 1970s which embrace the decision by the EEC 6 to enlarge the community and its common market, the accession of three new coastal member states – Denmark, the UK and Ireland – and the creation of a common framework for the management of the EEC 9's fisheries also mark the final flourish and eventual demise of distant water fishing in the North Atlantic. This chapter will therefore cover two key events in the narrative relating to the development of European fisheries policy: the first and most significant enlargement of the EEC in the early 1970s and the long drawn-out negotiations over the formulation of the Common Fisheries Policy (1976–1982).

It is pertinent at the outset to ask why, at a time when the focus of the European Community was on developing the common organisation of the market for its members' agricultural and industrial products, the all-consuming Common Agricultural Policy (CAP) and the upcoming accession talks, was it felt necessary to take on the task of creating a common fisheries policy that served the interests of a very peripheral economic activity and made only a small, though nutritionally valuable contribution to the Community's food security. There were some

DOI: 10.4324/9781003362913-4

for whom a common fisheries policy was the natural companion to the CAP and a requirement of the Treaty of Rome, though it is difficult to find any direct reference to this in the text of the Treaty. There were others who saw an opportunity to secure the benefits of determining the basic principles of Community fisheries policy in advance of any enlargement of the EEC, especially when such an enlargement might involve coastal states with substantial fishing interests. And there were a few who recognised the growing and urgent need to manage coastal fisheries that were coming under increasing pressure as a result of the expanding fishing effort. As the following accounts of early attempts to establish a common fisheries policy and the accession of Denmark, the UK and Ireland will make it abundantly clear, it was the second, opportunistic motivation that prevailed. In this and the subsequent development of the CFP my guide is Mark Wise's meticulously detailed *The Common Fisheries Policy of the European Community* (1984); the interpretation of events and their repercussions are my own.

Pre-accession manoeuvres

The prospect of enlarging the European Community beyond the six original signatory states of the Treaty of Rome gathered momentum throughout the late 1960s and with it came pressure from among the four Atlantic member states – and France in particular – to develop a legal framework for the application of fishing rights and the extension of common access to the territorial waters of present and future member states. Existing arrangements were based on the London Convention (1964) which recognised the 12 nm territorial waters of its signatory states and granted limited reciprocal access rights for fishing certain named species. The configuration of the EEC 6 provided little scope for a meaningful common approach to access rights and resource management, but the imminent prospect of enlargement made this a much more exciting proposition.

Responsibility for developing a Community approach to fisheries lay with the Commission's Directorate-General for Agriculture. Progress was initially slow and limited in content to the common organisation of the market and the principle of equal access that would dissolve the notion of national sovereignty as applied to territorial waters. Proposals published in 1968 were welcomed by the European Parliament but stalled in the Council of Ministers where contrasting objections from France and West Germany highlighted their very different profiles as fishing nations. However, the imminent approach of the date for opening talks with four would-be member states gave renewed urgency to the intention of having an *'acquis communautaire'* in place before their start. An agreement was finally reached on 30 June 1970 relating to three key areas: (i) equal access to all Community waters with time limited exceptions concerning the 3-mile limits in regions primarily dependant on fishing; (ii) the provision of structural funds over a three- to five-year period to assist adjustment to the new regime; and (iii) a decentralised approach involving Producer Organisations and the introduction of a withdrawal price mechanism to help stabilise market prices. Measures relating to the conservation of the fish stocks were, however, totally ignored. The concept of equal access had become the guiding principle of the yet unborn CFP and, of more immediate concern, the basis for the accession negotiations was laid. For the conspiracy theorists the trap had been set.

Enlargement

Within hours of these last-minute manoeuvres, negotiations were formally opened with four very diverse applicant states. It is important from the outset to recognise that within the overall scope of the talks fisheries were at best a minor concern, though the protracted negotiations over fisheries became an increasing source of irritation for the EEC 6 and the applicant states alike. A late-night conversation in the late 1990s with a retired senior civil servant, who at the time of the accession negotiations was a junior member of Edward Heath's negotiating team, left me with three clear impressions: nothing was going to get in the way of Heath's ambition to take the UK into the EEC; fishing interests came well down the list of the UK's priorities; and the UK's negotiating team had been seriously wrong footed by the *acquis communautaire* relating to equal access. Only in the case of Norway were fishing-related matters of any real significance to the overall outcome and it was largely, but not solely, on their account that ultimately Norway felt obliged, following a 53%–47% referendum vote against joining the EEC, to reject the Community's terms of entry.

Although the fisheries agenda contained two key items – the organisation of the market and equal access – it was the latter that dominated proceedings. Both Denmark's and the UK's negotiating positions were complicated by conflicting sets of interests in relation to fisheries. In the case of the UK, it was the aspirations of the then active distant water interests on the one hand and the domestic inshore fishing on the other. For Denmark, the division was between its domestic fishing sector whose interests were quite widely dispersed throughout the North and Baltic seas and the need to protect the interests of its semi-independent territories of the Faeroes and Greenland both harassed to some degree by distant water fleets. Ireland's position was quite different. It was at the time, a remote underdeveloped and peripheral part of Europe, anxious to share the benefits of a common market and European regional assistance with a fisheries sector greatly in need of modernisation. Its principal concern over fishing was to protect its 12 nm limits from exploitation by foreign vessels.

The Community's opening position, in line with the recently completed internal manoeuvres, was a fairly uncompromising one. It insisted on the adoption of the equal access principle for all EEC member states, with equal access arrangements phased in over the ten-year transition period with limited exemptions in respect of the Norwegian coast north of Trondheim, the Orkney and Shetland island groups in Scotland and the Faeroes and Greenland. These exemptions would be temporary with the EEC reserving the right to rescind them at the end of the transition period. These proposals were rejected by all four applicant countries who, while grudgingly conceding the principle of equal access in the case of Denmark, the UK and Ireland, were insistent on achieving a more or less permanent derogation for very much larger stretches of their coastlines. Over the next few months the familiar procedure of negotiation involving proposal, counterproposal and compromise, was skilfully directed by the Commission and less adeptly handled by some of the applicant states.

The final outcome set out in the Accession Treaty 1973 was, as the Commission had intended, a workable compromise (except in the case of Norway) that secured the principle of equal access, made concessions in relation to the designated areas of exemption and their future handling and established a ten-year transition period for its full implementation. It also confirmed the arrangements for the regulation of the market, involving the introduction of withdrawal price mechanisms and the establishment of Producer Organisations as the key agency.

Reviewing the accession agreement, Wise suggests that the Commission made rather more concessions than the three 'successful' applicant states: a case perhaps of 'smoke and mirrors' in so far as the adoption of equal access was of lasting value to the Community, whereas the concessions relating to 'exemptions' appeared to be of a more temporary nature and subject to future review by the Commission. Does the process summarised above suggest a conspiracy to defraud the applicants of their basic rights in relation to fisheries? The pre-accession manoeuvres certainly had the effect of wrong footing the applicant states who had much to lose and little to gain from having to concede 'equal access'. The Commission played a skilful hand by bidding up their initial demands in the knowledge that among the EEC 6 all but France would be happy to reach a compromise set well below the initial bargaining position. The game was played out to the Community's general satisfaction although they lost the big prize – Norway – on the way.

Did Britain make major sacrifices in relation to its fishing interests as the subsequent folklore would have it? At the time, probably not: they achieved what both key elements of their fishing interests had been looking for. In terms of what was to come, probably yes: for it paved the way for a CFP based on the premise of equal access in respect of the coastal states' 200 nm EEZs. But to argue that the EEC had a collective premonition of the cataclysmic events of 1976 a few years ahead would be to suggest that the Commission supped with the devil.

Enclosing the commons: the reverse of the coin

The impact of enclosing the commons on the member states of the EEC 9 was almost the exact opposite of what happened in the North Atlantic. Whereas it brought a feeling of relief and liberation to areas like the maritime provinces in Canada, Iceland and northern Norway, for the Community's coastal states it meant a sense of confinement and tension. The reasons are evident in the map (Figure 3.1) emphasising the sheer irregularity of Europe's coastline, the division of marine space into a series of semi-enclosed spaces (including the Baltic, North

FIGURE 3.1 The European Community's common pond (EC 9)

and Irish seas, the English Channel *inter alia*) shared by up to six coastal states, and the further fragmentation of those spaces into unequal exclusive fishing zones, many of them far too small to contain the fishing fleets of their respective countries.

Solving the problem of confined operational space was, on paper, quite simple and well within the framework of the nascent common policy. It involved merging the EEZs of its six Atlantic member states into a single Community EEZ within which the recently confirmed principle of equal access would apply. Ultimately that was what happened, though not before objections from several member states variously affected by such a course of action had been fully considered. As the Commission noted, the loss of access to North Atlantic fishing grounds meant a considerable deficit for the Community as a whole in terms of production, costs of converting highly capitalised fishing vessels for other duties, an increased community trade deficit in fish and fish products, redundancies *inter alia*, affecting all EEC coastal states except Ireland. For the EEC member states, the benefits of introducing EEZs were hugely discriminatory. The UK was by far the most handsomely provided for with, according to Commission estimates, around 60% of EEC fisheries resources contained within its notional fishing zone. Ironically West Germany and Belgium, which had also been highly reliant on distant water fishing activity, were the least well compensated having only very small exclusive fishing zones. At the time, most domestic fishing interests stretched well beyond the confines of what nominally became their own national fishing zones.

For the Commission, enclosing the commons in European waters was the catalyst for developing a more comprehensive common fisheries policy. Early in 1976, it prepared an outline of its thinking involving 'optimal exploitation of the biological resources of the Community zone' and 'equitable distribution of this limited resource', together with a coordinated approach to the establishment of EEZs in Community waters. Reflecting the scale of such a project, a separate DG Fisheries was created in April 1976 with around 50 staff. But the Commission's proposals concerning EEZs were not to be agreed without a fight. Following several abortive attempts to improve the Commission's revised proposals it was left to a late October informal meeting of Community foreign ministers in the Hague to outline an agreed approach to the designation of national EEZs, the Commission's role in negotiating access to Community waters for third party fishing interests and the possibility of adopting common measures 'when needed' for the conservation of resources within the 'common pond'. The Hague Resolution, as it became known also sought to placate UK and Irish concerns by insisting that account should be taken of the needs of areas particularly dependant on fisheries when applying future provisions of the CFP and confirming its support for the further development of Ireland's fishing industry allowing a doubling in size of its annual catch to 150,000 t.

Despite lingering opposition from Belgium and the Netherlands to the apparent 'sea grabbing' by the principal beneficiaries, an agreement was reached for member states to legislate for the establishment of 200-mile EEZs from 1st January 1977 and for the Community institutions to assume responsibility for access arrangements and the distribution of fishing rights within the common pond. The interpretation of those access and fishing rights was to trouble the

EEC 9 and its DG XIV (Fisheries) over the next six years of negotiation for the completion of the CFP.

It was never going to be easy finding a commonly acceptable solution to the Commission's quest for a fair and balanced Community fisheries policy that ensures the sustainability of the resource and provides an equitable distribution of fishing opportunities among its member states within an emerging new world order relating to ocean governance. In short it was seeking an experimental design for a previously unknown situation. Before looking in more detail at how the search to create a common fisheries policy was to unfold over the next six vexatious years, it would be wise to step back and try to get a better understanding of the scale of the challenge. To do this we need to take a leaf out of Kooiman et al.'s (2004) approach to fisheries governance and briefly examine both the EEC as a 'governing system' and the nature of the Community EEZ as 'the system to be governed'.

Fisheries governance

The governing system of the European Economic Community of nine, independent nation-states is made up of three key components: the *Commission* as the architect responsible for developing policy proposals while attempting to formulate a Community perspective; the *Council of Ministers*, comprising the appropriate ministers from each of the nine member states, acting as the sole decision maker with powers to sign off the Commission's proposals into EEC law, reject them outright or refer them back to the Commission for amendment or more thorough revision; and the *European Parliament*, which at the time, had no executive control over policy and served largely as an advisory body. This might appear to be a fairly simple institutional formula for governance.

However, in the case of fisheries, the policy process differs significantly from that applied to the majority of policy areas in a couple of important respects. First, the Commission has exclusive, rather than shared, competence over the framing of policy proposals in respect of the conservation of fish stocks; and second, Council decisions on fisheries are communicated as Regulations not Directives, implying no discretion for member states as to the form in which they are transposed into national law. Moreover, one overriding feature of the Community's governing system is that, at the time of negotiating the creation of a common fisheries policy, decisions were made on the basis of unanimity, not majority voting. There is no majority party directing policy nor even a ruling coalition of member states in theory or in practice, to steer decision-making in a particular direction, only seven independent member states each with something to win or lose depending on the outcome. Reaching compromise solutions was not going to be made any easier in an enlarged Community where one of its newest members was a political 'big hitter', a strongly independently minded island nation whose assumed natural authority had been challenged and its pride sorely hurt by the loss of significant fishing opportunities through the eviction of its distant water fleet from the North Atlantic. These factors go a long way to explaining the protracted process of reaching a final compromise decision over detailed policy proposals.

The system to be governed – the fisheries and fishing industries of the Community's EEZ – is diverse, dynamic and highly complex. The basic geography of Europe's seas is complicated both in respect of the irregular and fragmented nature of the boundaries between land and sea and in the network of ecosystems contained within the EEZ. As pointed out in Chapter 1, the living resources of the sea form a common property resource wherein it is difficult if not impossible to define ownership rights either by nationality or by individuals. Although UN-CLOS III (1982) was to define whose responsibility it was to manage the fisheries it did not substantially alter the common property nature of the resource.

Fish are by nature mobile; their life cycles often involve the crossing of national boundaries so that inevitably they become an international resource. For a number of commercially important species the adult feeding grounds – the legitimate target area for fishing activity – straddle the EEZs of two or more states, creating a 'shared stock' in need of collective management. To complicate matters still further, the feeding grounds of several different species may overlap so that fishing will involve the capture of different species forming what is known as a 'mixed fishery' which in an ideal world would be managed as a distinct entity rather than as several separate species.

Fish should not be thought of as a single commodity. In reality, the various species targeted by fishers have different uses and command different market values. The majority are intended for human consumption though these may vary quite considerably in terms of market price; others known as 'trash fish' in American usage, are destined solely for reduction into fish meal and oils at much lower prices; and a few – principally herring and mackerel – are caught either for food or reduction, depending largely on the volume of supply and the state of the market.

Over the years, the fishing industries of the seven EEC coastal states have developed quite strongly contrasting profiles based in part on their geographical location and on their fishing tradition, although all are focused primarily on servicing their own national seafood supply chains. Thus, the UK, West Germany and to a lesser extent Belgium, had developed a marked dependence on distant water fishing activity; France and the Netherlands had placed greater emphasis on the exploitation of fisheries in Europe's near and middle waters; and Denmark had invested quite heavily in catching and processing industrial species found mainly in the North Sea. The industries were also at different stages of development. While the Netherlands, France and Denmark were concerned to protect, expand and diversify their interests in European waters, the UK and Germany were preoccupied with rebuilding and restructuring their fishing industries as a result of their exclusion from distant waters. Ireland had aspirations to transform its underdeveloped, mainly inshore, fishing industry into a much larger, more modern and diversified sector of the national economy.

But, as far as the creation of a common fisheries policy was concerned, the critical feature of the system to be governed was the very impact of enclosing the Commons and establishing national EEZs. While the merging of the EEZs into a single 'common pond' was, in a sense, the *raison d'etre* for the CFP, the member states were still highly conscious of the very uneven distribution of benefits resulting from the creation of EEZs. The anxieties and tensions that became evident over the next six years, and beyond, very clearly reflect the conflicts of interest between protecting established interests and realising the benefits of the new order, together with lingering doubts as to whether the EEC institutions had it in their power to square the circles through the completion of the CFP. In brief, in both its spatial and temporal contexts the challenge facing the Community's decision-makers could not have been any greater.

Completing the Common Fisheries Policy: the final steps

Returning to the narrative describing the slow, torturous and sometimes tetchy progress towards completing a common framework for fisheries policy and, in particular, to the all-important issue of 'who gets what', we find the battle lines looking much as they did before the declaration of national EEZs and the assumption that for policy purposes they would be treated as a single 'common pond'. Agreement over the adoption of the EEZs had left the question of what happens next largely unresolved. On the one hand, the majority of coastal member states preferred the *status quo ante* approach supporting the equal access principle championed by the Commission. It was more in line with the freedom of movement principle that underpinned the common market and, to a degree, it reflected the realities of fish as a common property resource. On the other side of the divide were two new member states, the UK and Ireland, each of whom stood to gain considerably from a 'dominant preference' approach to the distribution of benefits accruing from the newly established EEZs. Their position, however, was difficult to argue on legal grounds given the earlier adoption of the equal access principle. It relied instead on notions of natural justice, the rationale of geographical proximity and the realities of the new world order in relation to ocean governance.

The Commission's comprehensive proposals for a framework regulation in 1978, clearly intended to provide as much continuity for Europe's fishing industries within a context of reduced fishing effort and a modified pragmatic approach to equal access. A system for distributing fishing opportunities on the basis of historic track records was to be the first major test of the relative strengths of the equal access and dominant preference positions. The proposal was accepted with little demure by the majority of member states but rejected outright by the UK who found no concessions to the idea of dominant preference nor any compensation for the recent loss of distant water fishing opportunities which had left a large deficit in the national seafood supply chain.

As an indication of their intentions to exert greater control over fishing activity within their EEZ, the UK government took unilateral action to extend the area of the Norway Pout box in the North Sea. The box, covering a large area to the south of the Shetland Islands and to the east of the Orkney Islands, had originally been established as a Community measure to protect juvenile haddock and whiting feeding grounds from the industrial fishing for Norway pout – a trashfish species – prosecuted by the Danes. A year later, with the Commission's approval, it was designated an autonomous national measure. The UK proposed to extend the area of the box eastwards incrementally by two degrees a year to the limits of its EEZ, thus reducing the Danish industrial catch by 220,000 tonnes a year in return for long-term gains in whiting (37,000 tonnes) and haddock (26,000 tonnes). Alternative proposals by the Commission for a much smaller extension together with the introduction of a 10% bycatch limit were rejected by the UK on grounds that it would prove less effective as a conservation measure, create substantial discards of bycatch species and incur higher costs of enforcement. In response, the Commission sought a judgement from the European Court of Justice which in 1979 ruled that the UK's actions were discriminatory and contrary to Community law, thus reaffirming the Community's 'governing system' as the overriding authority.

It would be wrong to regard the Norway Pout Box episode as a frivolous affair. It very clearly indicates the kinds of interplay between the Commission and the member states that were necessary to establish the hegemony of the Community's governing system in determining the shape of the future common fisheries policy. But there were other more fundamental and increasingly urgent matters to attend to concerning the proposed system for determining the distribution of fishing opportunities among the member states. Throughout the second half of the 1970s, the Commission had been allocating shares in the TACs proposed by ICES and moderated by the Commission on internal advice, in line with the historical track records for each member state. It appears that there was little serious dissent among the member states though there were strong objections to the rapid scaling down of the herring TAC following the virtual collapse of the North Sea stock during the 1970s. However, when the 1978 Commission proposals contained a recommendation to base future allocations on historic catch records with no apparent consideration for external circumstances nor for regional species needs, the minds of certain member states – most notably the UK – quickly became more sharply focused.

The distribution of fishing opportunities through fixed shares of the TACs was seen as the essential counterweight to equal access and of more immediate and substantive concern to the fishing industry allowing for continuity in the nature of fishing activity. Historic track records by their very nature paid no attention to either the impacts of losing distant fishing opportunities or to the promised support for fisheries-dependent areas. The Commission was persuaded to re-examine these issues and in 1980 came up with a number of corrective measures including a greater focus on six major species – cod, haddock, saithe, whiting, plaice and redfish (later adding mackerel) – and the use of 'cod equivalence'[1] as a means of equalising the value of allocations. Further considerations included reallocation of national quota to take account of 'third country losses' and the Hague resolution's transfers to assist designated fisheries dependent areas. Although the actual amounts of quota reallocated were quite small, the UK was clearly the principal beneficiary (third country compensations and special area benefits), along with Ireland (special area), West Germany and Belgium (both third country compensation) and the net losers Denmark, France and the Netherlands.

When the Commission published its final, virtually last-minute proposals for a common fisheries policy in October 1982, it proved to be a remarkably conciliatory document, incorporating the modified system for the allocation of TACs, recognising member states' exclusive rights in respect of the 0–12 nm limits with allowances made for the exercise of historic fishing rights in the 6–12 nm zone, and 'special status' accorded to North Britain, Ireland and Denmark's semi-independent territories of the Faeroes and Greenland. Moreover, it noted that the mechanism for allocating shares in the TACs 'assures each Member State relative stability of fishing activities for each of the stocks considered'; the concept of 'relative stability' was to become one of the defining principles of the CFP over the next 40 years. In addition, the Commission set out proposals to provide 250 m ECU over the next three years for measures to 'adjust capacity and improve productivity'. With minor amendments, these proposals were agreed by all member states with the single exception of Denmark, concerned in particular at the losses it incurred through the modified system of allocating TACs. Following further minor adjustments, *Council Regulation (EEC) No. 170/83... establishing a Community system for the conservation and management of fisheries resources* came into force on 25th January 1983: thus, after a lengthy and difficult gestation, the Common Fisheries Policy was finally born.

The comparatively smooth final stage (1980–1983) requires some further investigation. Was the generally conciliatory mood simply the result of 'negotiation fatigue' and the dawning realisation that cherished national ambitions were unlikely to find a home in a major Community project? Or was it the fear of a 'no deal' future for Europe's fisheries and fishing industries, the threat of anarchy and the possibility of multiple 'cod wars' in defence of national EEZs? Remember the fisheries deal could always be taken off the table without risking damage to other policy interests. Or did the completion of the CFP have a more immediate and tangible meaning? Was it better to seal the deal now in anticipation of the entry of Spain – a formidable fishing nation – and Portugal by creating a robust *'acquis communautaire'* in advance of potentially difficult negotiations, as the EEC 6 had done some 11 years previously? It is likely that each explanation had a role to play.

It is also interesting to consider the shift in the UK's position from *'agent provocateur'* to the much more amenable approach that occurred in or around 1980. In the 1970s it was the Labour Party that was more inclined towards Euroscepticism than the Conservatives. The Labour Party was in government from 1974 to 1979 with John Silkin holding the brief for fisheries negotiations – a tough, almost intransigent defender of the 'dominant preference' approach, close to the hearts of the UK fishing industry. The Labour government was replaced in 1979 by the Conservatives at the start of a 13-year period of office, mostly under the leadership of Margaret Thatcher who was to become a thorn in the side of the European Community during her later years in office. Was the clear change in approach adopted by Geoffrey Rippon and later Peter Walker as the UK's more mild-mannered representatives on the Council of Ministers simply a change of personal temperaments or a determination on the part of the Conservative government for Britain to be seen as a more compliant, responsible member of the EEC 9, or a realisation that always swimming against the tide was never going to get you very far? Or was it once again a case of using fisheries as a pawn in a much bigger game?

Significantly the ice had been broken in the summer of 1982, prior to the publication of the Commission's final proposals, when, following a series of bilateral discussions mainly involving the UK and France, a workable compromise was reached over the troublesome 0–12 nm 'territorial waters'. This involved the UK conceding its proposals for extinguishing all foreign fishing rights within the 0–12 nm zone while France and other member states relinquished their claims for equal access throughout the zone; in effect, this meant a return to the *status quo ante* established by the London Convention of 1964. Although of itself a fairly minor compromise within the greater scheme of a common fisheries policy, it probably helped to change the atmosphere surrounding the final round of negotiations.

Overall assessment

By way of conclusion, it is appropriate to ask whether, given the struggle over the preceding six-year period to reach an agreement on a common framework for managing the Community's fisheries, the journey had been worthwhile. Just how robust was the CFP likely to prove as a means of managing the future needs of the resource in terms of its sustainability and the diversity of fishing activity across the EEC 9? It is also reasonable to enquire how fair and reasonable was the system for distributing fishing opportunities between member states. Some may quite reasonably argue that it is best to judge a project like the CFP by its results rather than as a design – and we shall be doing just that throughout the next few chapters. But it is also still useful at this stage in the narrative to get some measure of what had been achieved as a result of the lengthy negotiations. Moreover, it will provide a sensible starting point for the next stage (Chapter 5).

Considering the scale of the challenge of matching the governing system with the system to be governed, the dramatically changing circumstances during the first half of the 1970s, the depth of feeling over the choice of direction of travel for Community policy, and the relatively short time available, the completion of COM (EEC) 170/1983 was little short of a miracle. In the space of a few years, we had moved from a situation of lightly regulated *open* access fisheries to an ostensibly more disciplined system of regulation and enforcement operating across the maritime boundaries of seven coastal states. The CFP was the first substantive expression of a permanent international approach to fisheries governance with the authority to enforce its collective decisions.

The system was, however, flawed; some of the flaws could be described as quite substantial and likely to prevent the Community from achieving its ambition of creating a fully comprehensive fisheries management system. A number of key issues stand out. Perhaps the most obvious and surprising for a policy framework intended to address issues of overexploitation and depleted resources was the absence of any clear strategy for managing the fish stocks; there was no clear statement of aims and goals, beyond ensuring 'by an appropriate policy for the protection of fishing grounds that stocks are conserved and reconstituted'. Instead, there appears to be a presumption that ICES through its annual advice on the state of the fish stocks would provide sufficient guidance and that strict application of the quota management system, supplemented by a range of technical conservation measures would be adequate for achieving sustainability. In short, it was possible to argue that the CFP offered more by way of a basic toolbox of measures for ensuring resource conservation than a clear and comprehensive strategy for achieving resource sustainability.

There was also an uncomfortable disconnect between resource management and the practical issues of fleet management – the former under the exclusive control of the Community institutions, the latter a devolved matter and therefore a member state responsibility. It is, however, impossible to exercise full control over fishing effort without having a considerable measure of influence over fleet management: the basic threat to the sustainability of fish stocks is the risk of overcapacity within the fishing fleet.

A division of responsibilities between the EEC institutions and the member states was clearly necessary and there were sound reasons for the decision to give exclusive power to the former over the management of fish stocks *per se*, for there could be no discretion allowed at the member state level in the formulation and implementation of conservation measures. But it would not be in keeping with the very nature of the Community to take control over all aspects of fisheries management. Decisions relating to the economic management of the fisheries were left in the hands of the member states most notably over the choice of an internal quota management system, fleet management and the inshore fisheries. The Commission would not have the resources nor the intimate knowledge of how fishing activities were organised across seven quite diverse fishing nations to be able to cope. Indeed, the Commission had already gone much further than in any other policy areas in assuming exclusive responsibility for determining the course of resource management.

Linked to the question of overcapacity is the issue of restructuring the fishing industry in the aftermath of enclosing the commons both in the North Atlantic and in European home waters. In modern times, the fishing industry is always in a state of restructuring in response to changing internal factors (production costs, market demand), new technologies and occasionally transformational changes within the political system. Post-war, the restructuring mainly involved the expansion and modernisation of the distant water fleet. In the late 1970s and early 1980s the focus was on rebalancing fishing capacity between the distant water and coastal segments of the fleet while trying to contain capacity expansion within the limits of a sustainable resource and a much more confined action space. As a separate decision the Commission had sanctioned the establishment of a restructuring fund for 1983–1986 and allocated 250 m ECU as a means of easing the transition to the new realities. But again, it offered little by way of guidance on how to prevent restructuring from turning into an expansion of fishing capacity.

And so, we come finally to the vexed question of equity in the brokering of the CFP and to the suggestion of conspiracy against the UK to deprive it of its proper share in the distribution of fishing opportunities. Was it the case that the trap, initially set by the agreement over 'equal access' as the guiding principle prior to the access negotiations and primed with the creation of a single Community zone through the merging of national EEZs in 1976, was finally sprung with the adoption of modified historic track records as the basis for allocating fishing opportunities? Certainly, there is some clear circumstantial evidence in support of such a claim. It was a simple matter of mathematics: how else could you explain the fact that the UK, through its EEZ, contributed something over 60% of the EEC 9's catch potential, should receive less than 40% of fishing opportunities? But consider it from the other side: the historic track records on which the allocation of future fishing opportunities was to be based gave a very accurate picture of how the UK was utilising the catch potentials of the seas around its coasts. And it was hardly likely that the Commission would wish to punish the remaining coastal states for the UK's loss of distant water fishing opportunities by abandoning the principle of equal access and destabilising their fishing industries. Perhaps we should leave the final judgement to Peter Walker, the UK fisheries minister during the latter stages of the negotiations, who in an article in *The Times* in January 1983 hailed the CFP as 'from a British point of view … a superb agreement'. However one chooses to judge the outcome – as a conspiracy to defraud or the inevitable outcome of difficult negotiations – a nagging sense of injustice was to haunt the UK fishing industry and colour its perception of the CFP for a long time to come.

Note

1 For the purpose of estimating the approximate aggregate value of member states' TAC allocations a system of 'cod equivalence' was calculated: cod, haddock and plaice (1.0), redfish (0.8), saithe (0.77), whiting (0.36) and mackerel (0.3).

Further reading

Wise, M. (1984). *The Common Fisheries Policy of the European Community.* London: Methuen.

Alternative, but less detailed perspectives are provided by:
Farnell, J., & Ellis, J. (1984). *In Search of a Common Fisheries Policy.* Aldershot: Gower.
Leigh, M. (1983). *European Integration and the Common Fisheries Policy.* Bekenham: Croom Helm.

4

CONTEXTUAL CHANGE

1970–2020

Introduction

Before resuming the narrative, it might be helpful to pause once again and to consider the complexity of the challenges in facing up to the task of managing a small but diverse sector of the economy that finds itself at the convergence of four vectors of change – economic, social, environmental and political. The latter has already been dealt with in some detail. This chapter will therefore explore the implications of economic, social and environmental change occurring over the period from 1970 to 2020 for both the behavioural activities of the fisheries sector and the development of fisheries policy.

Generally speaking, the pace of economic and social change appears to be increasing. But it is not just the pace of change that has intensified: the magnitude of the change has grown so that never before have the fisheries been exposed to such deep and far-reaching pressures. Superficially the situation is similar to that facing agriculture but with one crucial exception. In agriculture, science and technology have allowed farmers to specialise and intensify their means of production beyond the natural limits of the soil. Through the use of artificial fertilisers and pesticides, together with the breeding of resistant strains of crops (and livestock), agriculture has been permitted, at least temporarily, to escape the confines of the natural ecosystem. Marine capture fisheries by contrast remain almost completely dependent on the functioning of the natural environment and the productivity of the local marine ecosystems.

Marine aquaculture,[1] on the other hand, follows the principles of intensive farming allowing producers to transform the system of production by caging the fish, 'housing' them in abnormally large numbers, accelerating their growth rates through artificial feeds and controlling disease through pesticides, but also incurring some of the negative impacts of intensification mainly in the form of

DOI: 10.4324/9781003362913-5

environmental pollution. Globally marine aquaculture has made a considerable contribution to the production of seafood, mitigating the effects of decreasing levels of catch from marine capture fisheries. In European waters, its growth has been less remarkable and largely confined to the production of farmed salmon, mainly in Norway and Scotland, for both domestic and export markets.

The questions that the changing social, economic and environmental contexts pose for the present investigation are two-fold. The first concerns how far the basic frameworks for policy making are able to cope effectively with the pace and direction of external change. Where, for example, the pace of change exceeds the rate at which the governing system is able to respond, it usually means that policy making becomes preoccupied with solving yesterday's problems rather than planning for the future. The second question is whether the policy frameworks are sufficiently flexible to enable the fisheries sector to take appropriate adaptive actions to cope positively with the threats and opportunities that change inevitably implies. The answers to both questions will form part of the analysis of the first 40 years of the CFP in Chapter 5.

Economic change

Many of the economic changes affecting Europe's fishing industries in the later decades of the 20th century were a continuation of trends initiated much earlier, and now etched more deeply into the economic landscape. The processes of modernisation involving specialisation, economies of scale and the structural and spatial concentration of the fishing industry had led to fewer but larger fishing vessels, ports and quayside markets. The numbers of part-time or seasonal participants in the industry were being greatly reduced and fishing was becoming increasingly a full-time professional occupation. The pelagic sector is almost a perfect example. As late as the 1950s, the herring fishery had depended upon a myriad of small and medium-sized boats operating for relatively short winter and spring seasons throughout the coastal waters of the north east Atlantic. Today the fleet is reduced to a few very large mid-water trawlers based in a handful of ports fishing the region's offshore waters. It is technologically the most sophisticated segment of Europe's fishing fleets in terms of vessel design and fishing gear, but paradoxically active for as little as 65 days in the year.

To these established basic trends can be added two new elements of change, not altogether unrelated: the globalisation of trade in fish and fish products and the intervention of a neo-liberal market economy. The former helped the UK seafood supply chain adjust to the loss of distant water landings: in 1975 the UK was importing only 17% of its total supplies of fish for the domestic consumer market, but by 2017 the share of imported fish had risen to over 70%. Indeed, by the end of the 20th century, Britain had reached the seemingly incongruous situation where it was importing most of the fish it consumed and exporting most of the fish caught by its own fishing fleet. The price paid for fish on the quayside market was no longer determined by the abundance or scarcity of local landings but by the volume of fish available to a much wider, macro-regional market. Most of the imports into Britain were sourced within the north Atlantic basin and most of the exports of fish and fish products were destined for markets within the European Union.

The rise of a neo-liberal market economy within the fisheries sector, seen by many as the natural accomplice of globalisation, has over the last 30 years or so become a focus for debate within both political and academic circles. Broadly, it is concerned with the market having the freedom to intervene in the creation and distribution of wealth: in fisheries, this means the privatisation, valorisation and transferability of fishing entitlements in the form of individual transferable quota,[2] where the individual has the right to lease or sell all or parts of this quota to other fishers at the market rate. In theory, this private trade should help to achieve the optimal distribution of fishing opportunities, reduce wastage and so eliminate inefficiencies. Formal adoption of this mechanism has been limited to a comparatively small but growing number of fishing nations. In the North Atlantic, only two countries – Iceland, a pioneer in the development of the market economy in fisheries, and Denmark in 2007 – have embedded a system of ITQs in law codes governing fishing activity. However, in many other coastal states, the principles and practices of a market economy have been informally allowed to develop, with the UK a prime example. As we shall discover later this can lead to distortions in the pattern of quota ownership that are difficult to address. Generally speaking, fisheries economists tend to be supportive of the market economy seeing privatisation of fishing rights as the final clue to solving Hardin's tragedy of the commons dilemma and so promoting both a sustainable and efficient fisheries sector. By contrast, fisheries social scientists are usually critical of the privatisation of fishing rights mainly on the grounds of the accumulation of fishing entitlements in far fewer hands, the resulting squeeze on small-scale fishing interests and loss of employment especially in remoter parts of the rural periphery.

Social change

As with economic change, many of the causes of social change involve a continuation and deepening of processes that began much earlier. The dominant social trends of the late 20th century have been demographic change centred on falling fertility and resulting in the slowing down of population growth and increased social and spatial mobility especially among the young. In remoter rural areas the combination of these two trends has culminated in smaller, simpler nuclear households, the ageing of the resident population and depopulation. In addition, changes to the education system including the raising of the school leaving age, the broadening of the curriculum and the greater provision of technical education have all conspired to strengthen the likelihood of children from fisher households looking beyond the limited alternative employment opportunities in less privileged coastal peripheries and to seek their futures elsewhere. Among those who preferred to remain, some may find themselves driven out by the lack of affordable housing, especially in areas attractive to second home buyers and retirees.

These socio-demographic trends have had a major impact on the key social structures of coastal fisheries, weakening the form and functioning of the fisher household and the fishing community and undermining the traditional pluriactive economy of remoter rural areas. The combination of farming and seasonal or part-time fishing was a familiar mode of survival and petty commodity production flourishing throughout large areas of the Atlantic fringe from northern Norway to the south of Spain. Replacing the large, three generation extended family household – common throughout the Atlantic fringe in the first half of the 20th century – with the smaller, two generation nuclear household reduced the labour reserves available to work the family farm or manage the family fishing enterprise and threatened the integrity of the household as a viable work unit owning and operating the means of production.

Previously, the small family fishing enterprise had been crewed by members of the family household (father–son) or by other closely related kin (brother–brother; uncle–nephew), while the shore crew responsible for baiting the lines, mending the nets and preparing the fish for storage (drying, curing) or for sale was made up of the wives, daughters and retired members of the household. Larger coastal fishing enterprises with crews of four or more men usually sought to build the nucleus of their crews from within the extended kin network or the local community. With the emergence of the nuclear family as the norm, together with later school leaving age and the dispersal of younger members in pursuit of further education or alternative forms of employment, such close kin-based or local networks were greatly reduced. Crew recruitment became more difficult, and the social reproduction of the family-based fishing enterprise, based on succession and inheritance, was threatened.

Moreover, younger wives and grown-up daughters within the fisher household, having developed a wider skill set through improved education provision, were much more likely to look for regular employment beyond the limits of the family fishing enterprise. Regular incomes from non-fishing sources were likely to prove a welcome, sometimes crucial, supplement to what might otherwise be an uncertain and irregular revenue stream. The concept of a 'pluriactive' household had thus changed its meaning.

The identity and integrity of the fishing community have also been compromised. With an ageing population, fewer economically active households and a decline in the number of active fishers the links between many small coastal settlements and the fishing industry have been weakened. In some 'favoured' locations houses stand empty for much of the year being used as second homes or holiday lets; ties with the fishing industry have been further diluted by the in-movement of older people seeking an attractive place in which to retire. Surviving fisher households have, in some instances, become an 'encapsulated community', the term Ray Pahl coined in the 1960s, to describe residual agricultural populations in the so-called 'commuter villages' around the peripheries of large urban concentrations. To a degree, therefore, the 'fishing community' has become a mental construct based on historical association rather than a functional reality. The reality is that the fishing community has become a more diffuse and spatially dispersed entity.

At the same time, the physical landscapes of many coastal villages have altered. The harbour, once the focal point of the local fishing community, is no longer the exclusive domain of the fishing industry. Owners of larger fishing vessels, while preferring to remain living in the local community, have tended to move their operational base to the larger, better equipped urban fishing ports and closer to the quayside market. As a result, in the smaller harbours fishing boats may be greatly outnumbered by leisure craft and their quayside buildings converted from their original, fishing related functions to form part of the recreation/tourism infrastructure.

Environmental management

By comparison with the effects of economic and social change, the impacts of environmental change have so far been slight. However, the three main areas of concern – the impact of environmental legislation on the fishing industry, evidence of the growing power of 'environmentalism' to influence fisheries policy and the likely outcomes of the impending 'climate emergency' (a.k.a. global warming or climate change) – together create a dangerous cocktail of issues whose seeds are already sown and whose potentially catastrophic effects lie somewhere in the future.

A major ongoing problem is the increasing competition for space in the marine environment. Until the second half of the 20th-century fishing could claim a virtual monopoly of three-dimensional space use, but this has changed quite dramatically over the past 50 years, particularly in coastal waters where most fishing activity occurs. New uses have made quite an impact, including the exploration and development of non-renewable energy resources (oil and gas), renewable energy generation (wind farms, title energy schemes) – all of which require extensive exclusion zones – and the designation of marine protected areas (MPAs) for habitat and wildlife conservation. In recognition of the mounting pressures on marine space, most coastal states have over the past 30 years invested in the development of marine spatial plans, essentially broad strategic statements relating to the future management of marine space. Somewhat surprisingly, fishing interests often appear to be downplayed possibly as a result of naive assumptions concerning the fishing industry's ability to relocate elsewhere or, more likely, because of the industry's surprising reluctance to participate fully during the early phases of the consultations.

To date, environmental legislation mainly relating to the designation of MPAs has been a source of inconvenience, irritation and local despair at the perceived threat to fishing opportunities. Marine protected area is an umbrella term for a variety of designations ranging in scale from marine national parks to the myr-iad of much smaller protected sites. The protection they afford to endangered wildlife species and habitats is somewhat ambiguous ranging from the relatively rare 'no take zones', where all fishing activity is banned, to areas where certain types of gear are excluded either permanently or for specified periods of the year. In the long term, of course, MPAs can benefit local fishing interests – acting as refuges for endangered commercial species and helping to replenish depleted stocks. Current issues concerning MPAs relate to the lack of meaningful consul-tation with local fishing interests over their siting and the apparent lack of review procedures implying that their designation is permanent. But what concerns the fishing industry most is the rolling out of large MPA networks that would even-tually cover around a third of all coastal waters and uncertainty over the potential for MPAs to become significantly more restrictive in relation to fishing activity. Pressure has been building in environmental policy circles for MPAs to become more proactive in the protection they afford to habitats and wildlife at risk of future disturbance from fishing.

For some, the solution to the overexploitation of the living resources of the sea requires not simply a modest moderation of fishing effort but a dramatic restruc-turing of the way we use our seas. George Monbiot in his chapter on 'rewilding the sea' in his book *Feral: rewilding the land, sea and human life* (2013), skilfully draws attention to the true scale of our depletion of fisheries using the Shifting Baseline Syndrome to explain how we are deceived into believing that the ero-sion of the resource base is confined to the modern era. Each historical account of exploitation levels and their impacts on stock levels inevitably establishes its own baseline, measuring the depletion accordingly but ignoring the fact that it is simply an arbitrary point on a much longer timescale over which the decline may have been occurring. The 'success' of policy measures is often gauged by how close the catches and stocks get to that arbitrary baseline. To make the point Monbiot quotes the claims of the UK's National Ecosystem Assessment that 'UK finfish stocks are now at full reproductive capacity and harvested sustainably' as judged by the 1970 baseline by which date these fisheries had already been re-duced to a fraction of their full capacity.

In Monbiot's view, the state of UK fisheries calls for a rewilding of signifi-cant areas of our seas including the extension of no-take zones to around a third of our coastal waters, compared to a current estimate of 0.01%, the creation of buffer zones around many of the no-take zones, allowing only line fishing, and a universal ban on certain destructive types of fishing activity such as scallop dredging. Achieving such a dramatic transformation in the way we approach marine environmental conservation does seem unlikely, at least in the near future. But such an assertion ignores the possibility of mass mobilisation of public opinion in support of such moves. After all, the Royal Commission on Environ-mental Pollution's proposals in 2004 to dedicate up to 30% of UK waters to no-take zones was later endorsed by a petition bearing 500,000 signatures organised by a coalition of environmental organisations. The trend within Europe towards 'populist democracies' also supports the contention that all things may be possible, though admittedly the present political moves seem to point in the opposite direction to that followed by populist environmentalism.

Climate emergency

Looking further ahead, a much more formidable challenge will eventually confront Europe's fisheries sector. The emphasis will switch from finding basically simple solutions for unsustainable rates of exploitation to tackling more profound threats, triggered by global warming, to inherent capacities of marine environments to nurture and support the very ecosystems that presently sustain the commercial fisheries. The world's oceans, covering 70% of the globe, have in recent years become an important focus for climate change investigations. Sea surface temperatures have been rising since the late 19th century, most notably in the North Atlantic. Attention has been focused on the polar regions where the impacts of temperature rise on the thinning of the ice cap, increased fresh meltwater discharges and the retreat of sea ice have been observed over several decades. The likely impacts of global warming are generally well known: sea level rise, increased instability of weather patterns, increasing salinity and stratification of the oceans and the acidification of shallower seas.

The broad implications for fisheries are becoming clearer: marine finfish can adapt to the warming of the oceans in two ways: by altering their life cycle behaviour or by migrating. Fish will tend to mature at an earlier age and become smaller in size, with implications for harvest yields; or they can migrate in pursuit of optimal spawning and feeding conditions. Both developments are occurring in the North Atlantic where the southern boundary of the dominant cold water species (principally cod and haddock) has been moving northwards over the last 50 years and warm water species (John Dory, squid, red mullet *inter alia*) have become more abundant with the so-called 'tropicalisation' of temperate waters.

Translating such observations into long-term projections of the fisheries is a risky and seemingly thankless task – risky because the evidence base is insecure especially over the timing, location and rates of change and thankless because policy makers can only store the information, wait until the situation becomes clearer and then prepare for action. Nonetheless, a team of American-based scientists have modelled the 'maximum fisheries catch potentials' for more than 1,000 fish and invertebrate species for the period 2005–2055 and mapped the results that have been incorporated in the findings of the Intergovernmental Panel on Climate Change. The results, described by the authors as conservative, are daunting to say the least. They point to a major redistribution of the global catch potential with the most severe decreases of up to 50% of the 2005 potential in the Indo-Pacific region of the tropics, a zone that overlaps the areas of south and east Asia currently responsible for around 70% of the global marine catch. Compensatory increases in catch potential are expected in the high latitudes of the northern hemisphere, serving as refuge areas for northward migrating demersal species and enhanced by the expansion of open water in the sub-polar region in the wake of sea ice retreat. Secondary level changes will also affect temperate waters where catch potentials are expected to shift away from the continental shelf towards the deeper offshore areas, favouring large-scale fishing operations at the expense of smaller-scale fishing activity in inshore waters.

Precisely how this will impact on Europe's fisheries is still rather hazy. It is likely to affect the two key strands of Europe's seafood supply chains quite significantly. A major dislocation of established global trade patterns in fish and fish products is the most likely outcome and while most of this is regional in scope trade flows between Europe and south east Asia have been growing. A change of emphasis in the food security strategies of the developing countries most severely affected by global warming could well involve a switch towards 'food sovereignty' where declining catches are used primarily for domestic consumption rather than for export. The impact on fishing activity in the temperate zones seems certain to be less dramatic, though not necessarily less transformative. The northward migration of familiar demersal species in European waters is already becoming more pronounced and the centre of gravity for Europe's fishing industries is gradually moving northwards and it will continue to do so at an increasing pace. If the predictions concerning the displacement of coastal fishing activity to offshore waters in the temperate zones prove correct there will need to be a major reassessment of Europe's fishing potential, more cautious management of coastal fisheries and a thorough reappraisal of the EEZ concept.

But all this is in the future; there is little sign of any strategic thinking within policy making circles as to how we might deal with such situations. Our policy frameworks are all contained within a short timescale 'bubble'. The question, therefore, is *when* do we start to place future fisheries management firmly within the context of climate emergency? Underlying this question is an increasingly uneasy sense that we may have got our timings wrong. Recent observations of both Arctic and Antarctic regions suggest that the melting of the great ice caps is occurring more rapidly than conventional climate change forecasts have predicted. Satellite observations also indicate that the Arctic has already lost 40% of its summer sea ice cover and 70% of its volume since 1979. There is increasing concern over the structures and supplies of plankton in the Arctic and the role this may play in accelerating the processes of global warming. Phytoplankton not only produces more oxygen than the world's forests but also helps to sequester CO_2 more effectively when consumed by larger organisms and carried into the deeper waters.

Recent recalculations of how the warming of the seas may affect fish populations in the high latitudes of the North Atlantic have led to a further refinement of expected catch potentials. Norwegian research has suggested a boom-bust scenario for their northern waters with the initial warming of the seas encouraging early life cycle survival creating a strong surge in fish populations before the same waters eventually become too warm for the fish larvae to survive! Even within the favoured North Atlantic, the future of the next generation remains uncertain and insecure.

Responding to change

Leaving aside the issues relating to climate emergency, we now need to understand something of how the processes of economic and social change have impacted upon the lives, values and behavioural characteristics of fishers. In general, the past 40 years of socio-economic change described above have conspired to transform fishing as an economic activity, occupational choice and way of life: the traditional supporting structures of the fisher household and kinship networks have been weakened and partly replaced by new business-related networks. Fishing is no longer the inevitable source of employment for sons born into the fisher household but much more an occupation of choice. As a workplace the fishing boat still presents a physically demanding environment; the hours are long, but the workload has been eased by the introduction of new labour-saving technology. Some of the traditional skills and knowledge remain highly relevant but now need to be complemented by IT skills in the handling of fish location and gear-setting technologies.

For those intending to commit to a lifetime of fishing, the goal of being a skipper-owner and the status and prestige it still commands in the local community is most easily secured through the traditional route of family succession. For the new entrant, with little or no family connection to fishing, becoming a skipper-owner is expensive: the start-up costs – buying a basic but well found, second hand under 10 m boat and gear – will cost around £50,000 and much more if the purchase of licence and quota is involved. Thereafter, climbing the ladder by means of moving to a larger, better equipped vessel remains financially challenging and usually involves external assistance. Though the values of independence, self-reliance and autonomy of decision-making reliant on self-generated local ecological knowledge, that go along with skipper-ownership, still count for a lot, their meaning has been somewhat eroded as a result of the increasingly restrictive measures now applied in managing the fishery.

Fishing is of necessity becoming a more professional occupation, combining traditional skills and knowledge with a more entrepreneurial mindset and a greater understanding of modern business practice. In place of the largely self-sufficient family enterprise, the skipper-owner of a 15 m vessel finds themself relying on an external network of professional support services with an agent responsible for handling the sale of their catch, an accountant and a bank manager – who is likely to own much of the means of production – responsible for financial guidance and maybe a lawyer as key advisors. One of the more striking, and in some ways disturbing, features of the modern fishing industry is the pressure to constantly modernise the fishing vessel and its equipment, whether out of necessity or possibly vanity. With this comes the need to take out substantial loans to keep the enterprise afloat, and thus the need to maintain high levels of fishing activity in order to pay off the loan before earning an income for the skipper and crew.

There are, of course, alternative models. Small-scale fishing enterprises continue to dominate in terms of vessel numbers and, while some may conform to the new business model, others remain closer to the classic 'petty capitalist' model of operation – often fishing seasonally or part time, handling small and possibly irregular catches and self-reliant in terms of the harvesting and selling of fish and in the maintenance of their vessels. At the other end of the scale is the skipper-owned large-scale enterprise with annual turnovers in the several hundred thousand pounds bracket. Company ownership of fishing vessels, in the sense of the distant water fishing fleets, is a thing of the past. Some of the larger vessels may be owned in partnerships or family-controlled limited companies where the external services of sales agents, accountants and lawyers may be brought in-house – and this is a slow-growing trend in Europe.

It will be evident from the foregoing analysis that the fishing industry is constantly being forced to adapt to the changing conditions. In order to be able to adapt successfully, the industry must prove itself resilient armed with the ability to diversify its activities accordingly. One question that will be addressed in the next three chapters is how far modern fisheries management in the shape of the CFP and the member states' own management responsibilities has enabled or disabled this process of adaptation.

Notes

1 Not to be confused with mariculture where certain shellfish, notably oyster and mussels, are cultivated on improved beds with some restocking and relaying of the animals but are otherwise largely dependent on the normal functioning of the intertidal marine environment.
2 Almost all European countries have developed systems for the distribution of national shares of the TAC as fishing entitlements or quota normally made on an annual basis to individual fishers (or, more specifically, their vessels) in accordance with their historic track records. This is known as the quota management system (QMS).

Further reading

Monbiot, G. (2013). Chapter 13, Rewilding the sea in *Feral: Rewilding the Land, the Sea and Human Life*. London: Allen Lane.

Royal Commission on Environmental Pollution, 25th Report. (2004). *Turning the Tide: Assessing the Impact of Fisheries on the Marine Environment*. London: The Stationery Office.

The nature and content of two relevant EU sponsored research projects are described in:

Symes, D., Crean, K., & Phillipson, J. (1996). *Devolved and Regional Management Systems for Fisheries. Final Report*. Hull: University of Hull.

European Social Science Fisheries Network. (1997). *Final Report*. Hull: University of Hull.

Phillipson, J., Symes, D., & Salmi, P. (Eds.). (2015). Resilience and adaptation of fishing communities. The Special Issue of *Sociologia Ruralis*, 55 (3), 243–377. Also contains case studies of issues covered in Chapters 4 and 5.

See also:

Symes, D. (1998). Towards 2002: subsidiarity and the regionalisation of the Common Fisheries Policy in Gray, T.S. (ed.) *The Politics of Fishing*, pp 176–193. London: Palgrave Macmillan.

Symes, D. (2000). Integrated management: the implications of an ecosystem approach to fisheries management in Kaiser, M.J. and Groot, S.J. (eds.) *The Effects of Fishing on Non-Target Species and Habitats: Biological, Conservation and Socio-Economic Issues*, pp 366–382. Oxford: Blackwell Science.

Statistical data are mainly derived from MAFF (various dates). *Sea Fisheries Statistics*. London: HMSO.

PART II
The Common Fisheries Policy

5

THE DECADES OF MISSED OPPORTUNITY

1983–2002

Introduction

It is never going to be easy for a new policy to find immediate favour with its client populations and so prove an instant success. This is especially true when the underlying concept – a common framework for the management of a natural resource shared among seven neighbouring, independent states – is itself a radical, untried construct, when the process of delivering the policy is untested and when the means of ensuring effective monitoring, surveillance and compliance across the participating states are both underdeveloped and underfunded. Still more so when the policy makers' vision runs contrary to the expectations of the participating states and their fishing industries. Its chances of success are yet further compromised by the fact that the object of policy – wild capture fisheries – lies beyond the direct reach of management and where scientific understanding of their behavioural patterns is, at best, incomplete. Yet these are precisely the conditions facing the EC policy makers, the seven coastal member states and their fishing industries at the start of 1983.

With hindsight, it is tempting to argue that the CFP never really stood a fighting chance. The test of good policy management, however, lies in knowing how to respond to initial setbacks, how to identify their causes and how to repair the damage before it becomes endemic. We have clearly learned a great deal about how to manage fisheries – and, more particularly, the fishing industries – over the past 40 years. Sadly, we have not always been smart enough to apply the lessons promptly and effectively through vigorous policy reform.

A key to policy success lies in ensuring a coming together of all the major actors involved around shared objectives, agreed targets and mutual recognition of the mission's importance. This is unlikely to be in place from the outset but achieved through a gradual convergence of opinion around a common ambition

DOI: 10.4324/9781003362913-7

during the early years of policy development. From the start, however, the CFP was disabled by disagreement and inconsistency largely but not entirely as a hangover from the fractious six years of negotiation, yielding a compromise that satisfied no one. Even within the EC's governing institutions, there were conflicting agendas with the Commission striving to deliver a science-based conservation policy aimed at maintaining an assured and stable future for the Community's fishing industry and food supply system, while the Council of Ministers sought to protect the interests of their particular fishing industries by insisting on higher TACs. Member states remained suspicious of the EC's true intentions and of each other's ambitions. Meanwhile, the fishers' distrust of the underlying science, rejection of the chosen form of restraint imposed in pursuit of the EC's policy aims and fear for the future viability of their livelihoods grew with each attempt to ratchet up the level of restraint.

The next two chapters will map out the course of the CFP over its troubled early years, examine its failings and achievements and explore the attempts to alter the direction of travel and the institutional barriers that have slowed the pace of reform. These first four decades of the CFP fall neatly into two quite distinct phases. The original decision to establish a common fisheries policy was time limited, subject to review in 1992 and legally expiring in 2002: this phase of the CFP's development is perhaps best described as a time of missed opportunities. Its successor, also subject to decennial review, is examined in Chapter 6 under the title of 'Changing Directions'.

In the preparation of these two chapters, I warmly welcome two major 'biographies' of the CFP – by Michael Holden and Ernesto Penas de Lado, both distinguished 'insiders' whose insights into the functioning of the CFP are hugely informative. They offer sharply contrasting accounts of the Policy, its aims, content and operational modes. Holden's *The Common Fisheries Policy: Origin, Evaluation and Future* (1994) is a passionate critique of flawed design and failing implementation in the first ten years, ending with a plea for a fundamental change of approach to the management of Europe's fisheries. Penas de Lado's *The Common Fisheries Policy: The Quest for Sustainability* (2016) spans the period up to 2015: his textbook-like approach is in essence a dispassionate and less judgemental, clinical account of the difficulties faced in the early years and the consequences of a change of direction taken in 2003 and strengthened in the latest Policy review. Together these two volumes provide a very useful framework for my own reflections resulting from much more limited engagement with the CFP over the 20 years from 1993 to 2012.

Bedding in: 1983–1992

The Common Fisheries Policy comprises four distinct pillars: the *common organisation* of the market in fish and fish products (Regulation 2140/70); *structural policy* first defined in 1970 with reference to 'rational development of the fishing industry within the framework of economic growth and social progress'

(2147/70) and today embedded in the share of the structural funds dedicated to the development of the fisheries sector; *external relations* initially formulated in 1977 to negotiate limited access for Community vessels to traditional distant water grounds but later focusing on access agreements with mainly African countries used primarily by the Spanish distant water fleet; and finally, *resource conservation*, the main focus of the negotiations from 1976 to 1982 and the core of the first Framework Regulation (170/83). Both the market and external policy pillars are generally regarded as qualified successes. In practice, when discussing the CFP what is usually being referred to is the contentious and – in some quarters – deeply unpopular resource conservation policy. And it is this element of the CFP that will dominate the following analysis.

Conservation policy: the role of TACs

During the first few years of the CFP, the primary concern was simply to accustom the industry to the implications of 'fishing within defined limits'. It was a period in which the objectives, mechanisms and tools for managing the Community's fisheries were being tested through practical experience. These early years proved an inauspicious start for the new regime – unsurprisingly since the conservation policy had been largely scrambled together through trade-offs and compromise with the aid of financial incentives. But the extent of the policy's incoherence and the lack of overall control and enforcement was disturbing. What was to prove truly alarming by the end of the decade, however, was the failure to deal effectively with the fundamental deficiencies despite the growing evidence of the damage being done to the resource itself.

What was exposed during the trial period were the flaws in the structure and content of a conservation policy where the objectives lacked clarity and were further obscured by a lack of cohesion between the different strands of its policy, and where the policy process laid itself open to what is known as implementation drift. Consider, for example, what happened in setting the annual TACs – the principal mechanism for assuring the health of the most important commercial fish stocks, designed ostensibly to reduce the pressure from fishing activity. Initially, as few as nine species were subject to so-called 'analytical TACs' involving scientific assessment and the calculation of safe allowable levels of catch, leaving the vast majority of species (but a minority of the actual catch value) outside the direct reach of the CFP's conservation policy. An increasing number of species were to be brought under the purview of the TAC process in the years ahead.

The initial scientific advice from ICES through its Advisory Committee on Fisheries Management (ACFM) was, in effect, reappraised by the Commission's own panel of experts in the light of the commitments of the CFP and in particular the provisions of the Hague Agreement. As a result, the Commission's TAC proposals usually, but not in all instances, were set at slightly higher levels than ACFM's recommendations. These were then put before the Council of Ministers' December meeting for decision. Unlike the Commission, which treats the setting of TACs as a science-based exercise, the Council sees it as a political obligation to secure the best possible results for its own fishing interests: hence the frequent description of the process as 'rampant horse trading' and the almost inevitable outcome of 'agreed TACs' clearly in excess of ACFM's scientific advice.

The second phase concerns the implementation of these agreed TACs by the member states through quota management systems (QMS) using a running sequence of individual vessel track records, with each member state free to devise its own quota management system as we have now crossed the boundary between the Community's and the member states' areas of responsibility. When it comes to the recorded landings, once again the results are at variance with the agreed TACs, usually above the Community's target level but surprisingly sometimes below not only the Community's target but also the ACFM's original advice. Exceeding the TAC can readily be explained by a combination of the fishers' anxiety to make the most of the limited opportunity and fear of losing part of their quota entitlement in future years, poor monitoring and weak surveillance and enforcement. The situation where recorded landings fall short of the ACFM's recommendation is less easily explained. The only viable interpretation would seem to be that ICES/ACFM's own assessments were flawed, overestimating the available adult populations and, as a result, creating so-called 'paper fish'.

Sadly, the evidence of increasing pressure on already declining stocks as depicted by official landing statistics tell only half the story. Official landing statistics only record legal landings. They cannot take account of illegal, unregulated and unreported (IUU) fishing which was on the rise as attempts to regulate fishing activity began to take hold. Nor do they include the amounts of fish that are discarded between capture and landing. Discards take two forms: high grading of the catch in order to retain the most marketable fish (in terms of species, size and freshness) and so command the best prices at the quayside markets; and the increasingly institutionalised practice of discarding in order to remain within legal catch limits as defined by the vessel's licence and catch quota. This was to become a major issue in succeeding decades especially in the context of mixed fisheries when the EU's continuing inactivity over discards was making a mockery of the very notions of sustainable fishing and achieving optimal economic returns.

Technical conservation measures

Although TACs bore the brunt of conservation management they were in theory supported by an array of technical conservation measures (TCMs) that were delineated not in the main Framework Regulation (170/83) but in supplementary regulations (171/83). They covered a range of measures – minimum landing sizes, gear regulations (that were intended to prevent the capture of fish that had not yet reached spawning status) and closed areas that served a variety of purposes of which protecting juvenile feeding grounds and spawning grounds were the most important. In practice, however, closed areas were used only sparingly. They were potentially discriminatory, reducing freedom of access for certain types of fishing and for those member states that traditionally fished those grounds. Moreover, along with regional variations in gear regulations, they disturbed the notion of a level playing field on which the concept of the CFP rested.

Managing fleet capacity: the role of MAGPs

It soon became clear that with or without the aid of TCMs, TACs were unlikely to prove effective in delivering the conservation goal set by the terms of the Framework Regulation. Without the means of controlling the overall harvesting capacities of the Community's fishing fleets, which had been growing significantly through new vessel construction and the remarkable advances in fishing technology, the conservation policy was destined to fail.

The idea of Multi-Annual Guidance Programmes or MAGPs had figured briefly in the original structural policy regulation (2140/70) as a means of coordinating structural policy measures and enabling appropriate financial assistance for the development of the fisheries sector that at the time meant improving fishing capacity. Although regulation of fleet capacity did not feature prominently in the first Framework Regulation (170/83) setting out the aims and objectives of the CFP, the concept of MAGPs was resurrected in Regulation 1908/83 with the aim of achieving 'a satisfactory balance between the fishing capacity to be deployed … and the stocks expected to be available'. The first of four sequential Programmes was inaugurated in 1983. Once again, the Commission tiptoed into the fray with very modest targets, unwilling to upset member states still recovering from the bruising encounter over the establishment of the CFP. This time it was with good reason, for there was no actual way of measuring fishing capacity. Two surrogate measurements were used – gross registered tonnage (GRT) and engine power (kW) – both well-established indicators of the size of the fleets but affording little or no indication of their harvesting potentials since neither took account of the increasing size and technical sophistication of the gears deployed.

Despite the extremely modest targets set, suggesting that the most the Commission was hoping to achieve was to contain fishing capacity at or about the existing level while signalling a change of approach to conservation policy, MAGP I (1983–1986) proved a sorry failure. All but two member states fell short of their targets and the EC's overall fishing capacity continued to grow. Part of the explanation may lie in the governing system's inexperience in handling such a scheme and the member states' unpreparedness, inability and lack of will to exert control. But the most formidable obstacle was the Community's own funding arrangements for the development and modernisation of the fishing industry, including grants for new vessel construction – further evidence of the incoherent nature of the CFP and the urgent need for remedy.

MAGP II (1987–1991) showed little overall improvement. Although attempts were made at more precise targeting of the fishing fleets in terms of species, *métiers* and fishing grounds, the new targets were still set well below those required to bring about effective capacity reduction: a 3% reduction in GRT and 2% for kW. Only four Atlantic member states (Denmark, Germany, Spain and Portugal) met their targets, while France and Ireland came close; three member states (Belgium, the Netherlands and UK) failed to reach their targets by a margin of at least 11% (Holden, 1994). The setting of targets and their achievement continued to be compromised by the lure of Community grants for new construction and the fact that many of the decommissioned vessels, which theoretically created the space for new builds, were old, inefficient and in some instances inactive. The persistence of increasing fleet capacity made it all the more difficult for the Commission and the Council to set lower TACs in the face of increased pressure to satisfy the needs of fishing fleets whose capacities were inexorably rising rather than receding.

Control and enforcement

So far we have paid little attention to the capabilities of the EC and its member states concerning monitoring, surveillance and control and the competence of their enforcement systems that are needed to underpin the effective delivery of fisheries policy. These systems varied in quality throughout the Community and were, in general weak – again reflecting the unpreparedness of member states' institutions for the task of regulating the fisheries. Official data sets were in some cases incomplete; inspection and surveillance systems rudimentary; enforcement capabilities both at sea and on land inadequate. Left largely to their own devices, member state bureaucracies were highly sceptical of each other's motives and competence. As a result, the implementation of the regulatory mechanisms of the EC's conservation policy was being seriously compromised by a combination of inefficiency and weak commitment.

Meanwhile, what was happening to the fish?

The true measure of how well a conservation policy is performing is its impact on fish stocks. While it may be academically unsound to limit the evidence to a single species, the case of North Sea cod offers a sufficient and appropriate illustration. After all, cod is an iconic species, as close to being a staple food as any species of fish could be and a traditional mainstay for the Community's fishing industries; and North Sea fisheries were a focal point for the EC 9's deliberations in the late 1970s and early 1980s.

The profile of ICES' assessments of total allowable catches for North Sea cod indicates that the initial downturn in stocks occurred around 1970 with its steepest decline in the 1970s and a slightly less severe rate of depletion beginning around 1983 but continuing throughout the 1990s and into the early years of the present century. The inference is therefore that the CFP cannot be held responsible in any way for the initial decline in the 1970s. However, after its full inception in 1983, it may have contributed to the slowing down of depletion, but it certainly failed to arrest it. ACFM advice was persistently urging stronger cuts in North Sea TACs; the EC was obstinately resisting this advice in favour of more liberal TACs; and recorded landings were consistently falling below the agreed TACs.

We must be careful, however, not to place all of the blame on the policy makers or the fishing industry. There are grounds for believing that in the aftermath of a long-term fluctuation in the environmental conditions in the North Atlantic, that had given rise to the 'gadoid outburst', annual recruitment of cod and kindred species to the spawning stocks was particularly weak. Nonetheless, the warnings were clear: the North Sea cod fishery was edging ever closer to the brink. During the next decade it was to reach the edge.

More of the same: 1993–2002

The new Regulation

A mid-term review of the CFP was required under the terms of the original Framework Regulation, presumably intended to pave the way for any necessary revision of the Policy during the remaining ten years of its legal lifetime. The Commission duly obliged with the publication of its Report 1991 (SAC 91/2208), a fairly comprehensive analysis of the state of the EC's fish stocks and of the measures deployed to ensure their protection, couched, however, in 'non-alarmist' language. It stressed the need to reduce fishing pressure on certain key stocks through better control of fishing capacity, tempered by the need to consider the economic and social consequences. It also drew attention to the impact of discards and the need for a multi-species approach to fisheries management. It was critical of the weak level of control over the implementation of policy measures while carefully avoiding any direct criticism of the member states' responsibility for monitoring, surveillance and enforcement of regulations. Overall, Report 1991 was a calm, contemplative and perceptive review, but lacked a sense of urgency and direction. It stopped short of making firm recommendations for reform; this and its deliberately soft language set something of a template for future decennial reviews that would reiterate the same basic failings of the CFP.

The new Framework Regulation (345/92) followed suit. Essentially, it confirmed and consolidated the role, objectives and competence of the original vision of the CFP by reaffirming relative stability as its key objective and continuing the unyielding reliance on TACs and TCMs to deliver its conservation policy and by leaving the dysfunctional fleet capacity mechanisms virtually unchanged. But it did include some surprisingly innovative thinking in relation to the possible introduction of a multi-species, multi-annual approach to fisheries management; moreover, it ensured the inclusion of economic issues within the revised remit of the Commission's influential scientific committee, renamed the Scientific, Technical and Economic Committee for Fisheries (STECF).

On the downside, the new Regulation ignored the crucial issue of discards and omitted any reference to the significance of the links with the other three pillars of a common fisheries policy, namely structural, market and external affairs. This cautious, unintegrated, conservative approach was later further amplified by the seemingly wilful neglect of the opportunity to move forward with potentially transformative ideas of multi-species and multi-annual management. The continuing disarticulation of the fisheries policy and the isolation of stock conservation from more or less all other areas of Community policy were both baffling and disturbing.

So what can explain this apparent 'isolationism' and more importantly the reluctance to engage more firmly with the key issues of fisheries management? Was it related in any way to the almost unique 'exclusive competence' of the Commission's role in respect of the conservation of 'the living resources of the sea'? Was it simply the perceived fragility of the CFP that turned the Commission away from challenging the member states through more fundamental reform – leaving the preservation of the *status quo* and the survival of the CFP as the only achievable goals? Or were there instructions from the Commission's inner sanctum not to risk 'rocking the boat' of the European project by insisting on more effective action to conserve fish stocks?

Whatever the cause, the most damning legacy of this first review and renewal of the CFP as initially conceived in 1983 is that it creates the conditions for what is known as 'path dependency', a situation in which policy development is largely limited to incremental changes based on the adoption and embedding of principles, objectives and mechanisms from the previous policy decisions in the prevailing policy culture. As a consequence, more fundamental changes of direction are ruled out of order and only modifications to the existing policy are negotiable. In the case of the CFP, this path dependency was strengthened by the addiction of most member states to relative stability and further intensified by the institutional inertia of the Community's governing system in later years. Given the alleged fragility of the CFP and the reluctance of member states to reengage in potentially fractious negotiations, it was not surprising that the Commission opted for the easier option of inactivity. But, as we shall see in Chapter 7, this made the adoption of innovative development in the early 21st century all the more difficult even though in DG Mare (as it became known), a majority of member states and key sectors of the fishing industry were quite receptive.

Outcomes

The overall outcomes of the new Regulation were largely inevitable, and the details need not detain us long. In summary, they involved a continuation of trends from the previous decade: an inexorable drift towards a crisis framed by severely depleted stocks and centred on a failing, if not failed, approach to managing the Community's fisheries. The scope for bolder action by the Commission and for greater pressure on the Council of Ministers to behave more responsibly in relation to the urgent need to reduce fishing effort had been denied.

Turning again to our earlier example of North Sea cod to provide some indication of the effects of what seems like woefully weak management, while the ACFM was urging the Commission to cut the TAC by up to 30% throughout the earlier part of the decade the agreed TACs for North Sea cod were persistently set very much higher. Some respite was afforded through good stock recruitment in the middle part of the period allowing ACFM perhaps unwisely to relax its insistence on high levels of cuts to the TACs for 1996, 1997 and 1998. In these years the Commission's agreed TACs were actually lower than those recommended by ACFM. But this brief recovery proved to be a false dawn. In the last two years of Regulation 170/83's life span the ACFM were counselling 'lowest possible catch' and preferably a zero TAC for cod.

It is only fair, at this point, to explain why it proved so difficult – indeed impossible – for the Commission to follow ACFM's advice. Cod lay at the centre of the all-important mixed demersal fishery in the North Sea that also included haddock, saithe and whiting, each targeted to some degree by the six member states bordering the regional sea, plus Norway outside the EC. To ban the catching of cod would thus be tantamount to closing down the entire mixed fishery, effectively crippling the region's fishing industries. And that was politically untenable. The best that could be hoped for would have been a much closer – though not complete – alignment with the scientific advice thus slowing the rate of resource depletion, but not halting it. To this day, the issue of mixed fisheries remains an unsolved problem of fisheries management.

Amid the gloom and despondency concerning the EC's attempt at conservation policy, there were some brighter intervals of realism, innovative thinking and movement towards greater convergence with conservation needs, manifested in the revised agendas for fleet capacity management, though sadly not reflected in the actual targets set. MAGP III (1992–1996) was set against the background of the devastating Gulland report (1990) that had estimated excess capacity within the EC fleet at around 40% and urged the Commission to adopt a stepwise regression in fleet capacity to bring the threat from overfishing under control. Ambitious objectives were to be set by the Commission and alternative means of attaining them identified. In response, the Commission did set more challenging targets of a 20% reduction for demersal roundfish fisheries and 15% for benthic fisheries. Remarkably by 1996, those targets had been met by all Atlantic member states except the UK and the Netherlands – an indication perhaps of much greater awareness among member states' governments and fishing industries of the disastrous long-term consequences of failing to rein in fishing effort.

So far within this analysis of the CFP little reference has been made to the roles of the member states acting within their own legal capacities. Member states are not, as some believe, simply ciphers of a highly centralised bureaucracy, slavishly translating EC regulations into national law. Fleet capacity reduction is a case in point where it is the role of the EC to set the objectives for capacity reduction, nominally in consultation with the member states, but the responsibility of the member states to devise their own independent strategies for achieving them. Here, one can sense that some member states are more inclined to ensure the implementation of EU regulations in both the letter and spirit of the law, others perhaps less so. There are other areas of fisheries management where member states have full responsibility for both policy formulation and implementation: quota management systems and inshore fisheries are the two most important.

The last of the multi-annual programmes (MAGP IV, 1997–2001) was also guided by an independent report on overcapacity (the Lassen Report, 1996) which pointed to much the same conclusions as the Gulland Report but proposed a somewhat different approach. This involved a more detailed segmentation of the fleet, a system of targeting that gets closer to the realities of existing fishing and reductions proportional to levels of actual overexploitation of the resource. But the Commission's response was again too modest, as evidenced by the fact that most member states had met their targets by mid-term. In the meantime, concerns were being voiced as to the true impact of the most recent programmes and the role of days at sea restrictions as an alternative means of attaining the targets – where, in fact, what was being achieved was not so much a reduction in capacity but simply a temporary relaxation of fishing effort achieved mainly through days at sea rationing. This was further highlighted by the decision to provide additional funding for the temporary laying up of vessels. Partly for these reasons, together with the unpopularity of the Programmes among most member states, the decision was taken to abandon them in favour of a much simpler but no more effective scheme involving the replacement of a decommissioned vessel by a new build with a set reduction in fishing capacity as measured by vessel length.

New ideas

Around the middle of the 1990s at a time when the EC and member states were struggling to implement the CFP in an efficient and effective way, academics were starting to show greater interest in issues relating to fisheries and their management. Fisheries science was already an established research field: fisheries economics likewise but on a much smaller scale. A new area of research was beginning to emerge: fisheries social science, a broad church of disciplines ranging from social anthropology and sociology through social geography and social history to the political sciences, intrigued by the impacts of modernising the fishing industry on fishing communities and the unique nature of fisheries as a natural resource and the challenges of its management. At the time, within Europe fisheries social science was to a large extent a diaspora of individual researchers lacking a coherent institutional identity. The challenge, generated by the crisis in the region's fisheries and their management, was to prove a catalyst for the development of the research field.

At this point, I have singled out four key themes in which fisheries social science had an interest. Each is related to management; each is concerned with an issue that had already taken or was about to take on a measure of significance in the delivery of the CFP; and each implies criticism in the way the Policy was being conducted. It is important, therefore, to recognise that DG Fisheries, far from being unreceptive to critical analysis and the promotion of innovative thinking, was in fact proactive in initiating certain new lines of inquiry. Indeed, certain of the research findings outlined below emerged from two innovative social science projects, focusing on fisheries management issues – sponsored under the European Community's AIR and FAIR research programmes.

A persistent and almost universal complaint of the fishing industry is that modern management systems rarely find space for fisher participation in either the framing or implementation of policy. The resulting lack of local knowledge and practical experience, it is argued, may go some way towards explaining why policies fail. While fisheries scientists may know something of the makeup and behaviours of fish stocks, bureaucrats and politicians have little or no understanding of fishers' values and behavioural patterns. Hence the impracticalities of some policy measures, the lack of commitment from the industry and a tendency to weak levels of compliance.

The theme of *co-management*, where government bodies and fishers' organisations share responsibility for the implementation of policy at the local scale, was initially explored by Norwegian and North American sociologists in the 1980s, focusing on their structures, remits, memberships and modes of operation. Subsequent investigations over a wider geographical area found abundant evidence of successful, long-established local co-management practice throughout Europe including *cofradias* in Spain, *prud'homies* in France and the remarkable network of Sea Fisheries Committees created in the late 19th century and still responsible for inshore management in England and Wales at the turn of the 20th century. Such systems usually operated in the context of small-scale inshore fisheries, but more modern applications of co-management were also to be found in the Dutch *Biesheuvel* system of quota management for larger vessels and in the evolving role of fish producer organisations (FPOs) – originally set up by the Commission in 1971 to assist the stabilisation of prices at the quayside markets – in managing the offshore sector. By the end of the 1990s, FPOs were handling around 95% of the UK's quota allocation. What remained missing, however, was the engagement of the industry representatives in the framing of fisheries policy at the European level. The emphasis on 'participative governance' was to become greater in the next stages of the CFP reform.

In some ways closely related to co-management initiatives but of greater concern for the CFP's struggle to find an appropriate *modus operandi* for its expanding empire was the theme of *regionalisation*. The CFP had successfully negotiated two significant extensions to the European Community – the first to the South in 1986 with the accession of Spain and Portugal; the second in 1995 with the inclusion of Sweden and Finland. As a result, the Community's jurisdiction extended over a huge, highly diversified area of maritime space, extending through 60 degrees of latitude. It was becoming difficult to justify the concept of the CFP catering for a single management area. Moreover, the geography of the 'common pond', fragmented into distinctive regional seas with different personalities and management needs, called out for a regional differentiation.

The first coordinated fisheries social science research project funded by the EC involved partners from Denmark, Spain and the UK, together with Norway, and was initiated in 1993 to explore the potentials for 'devolved and regional management systems for fisheries'. It involved the comparative analysis of the complex structures and relations involved in the delivery of fisheries management in the four countries and their potentials for further devolved responsibilities with a view to developing more effective management systems for both the EC and its member states. In its later phases, the project focused attention on the idea of regionalising the CFP, allowing the closer correlation of the aims and objectives of the CFP with the specific characteristics of fishing activity in the regional seas that made up the 'common pond'. Among the recommendations was the suggestion that the EC might wish to consider the idea of implementing a regional seas approach using the report's concept of Regional Fisheries Councils (Symes et al., 1996).

Unlike co-management and regionalisation, the two remaining themes were in no sense the property of fisheries social science. Both were initiated by other disciplinary interests. In the case of an *ecosystems based approach to fisheries management* (EBAFM) the role of fisheries social science was largely limited to that of advocacy in the interests of a more broadly based and fully integrated approach where currently the conservation of fish stocks appeared to be the sole driving force to the detriment of economic, social and environmental considerations. The single species approach had clear limitations, most clearly evident in respect of mixed fisheries. EBAFM pointed the way to a more balanced alternative, combining not only long-term security for commercial fish stocks but also the continuing health of the ecosystems that nurtured them.

Two different 'messages' emanated from EBAFM. The first was simply the need to avoid damaging or in rare instances destroying the marine ecosystems through fishing activity. Positive action included the protection of endangered habitats and species through the designation of marine protected areas (MPAs) – an easy enough, though still highly controversial, task and the responsibility of DG Environment and member states rather than DG Fisheries. The second, much more difficult 'message' was addressed directly to those responsible for fisheries management. It concerned the opportunity for a fundamentally different approach based on a deeper understanding of marine ecosystems and the interactions of fish – both between species and in relation to the broader ecosystem – that might allow policy makers and managers to place less emphasis on the blunt instrument of TACs in favour of more nuanced use of technical conservation measures to manage the stocks.

The difficulty lay in operationalising EBAFM – the setting of clearly specified objectives, the choice of key indicators and the formulation of practical ecosystem-based management plans. This still remains a largely unfulfilled ambition. Interestingly, the 2002 basic Regulation laid claim in Article 2 to the incremental application of EBAFM – a false claim for no ecosystem-based approach would tolerate the extensive damage to marine ecosystems arising from discards, but an indication of the growing need to acknowledge the influence of ecosystem interests.

In the final theme, which was to develop significantly in the ensuing decades, it was a case where the fisheries social science role was to counsel caution rather than enthusiastic support for its inclusion within the EU's approach to managing the fisheries. It concerned the introduction of *market-related systems of quota management* involving the use of individual transferable quotas (ITQs) that could be traded (leased or sold) as private property. As such it was a matter for the member states rather than the European Commission. This particular application of rights-based management, popular among fisheries economists, had worked effectively in both Iceland and New Zealand creating a more economically efficient system of balancing fishing capacity and fishing opportunities, leading to a restructuring of the fishing industries. But there were significant social consequences around issues of employment, the decline of small-scale fisheries and the future viability of established coastal communities.

The market approach was of particular interest to the EU governing system as it offered a major improvement on its current fleet capacity management scheme. But it was also of great concern for member states searching for the delicate balance between economic efficiency and the needs of small-scale fisheries and fishing communities in the remoter fisheries dependent coastal regions. At the time, no member state was willing to embrace the ITQ system formally, though in certain coastal states elements of the approach were allowed to establish a firm foothold. Pressures for its formal adoption were to become stronger in the years ahead.

Analysis

So what is one to make of the initial incarnation of the CFP and what were the prospects for its future? Who better to make an informed judgement than someone who had worked on both the formulation and early years of the CFP and served as head of DG Fisheries' conservation unit from 1983 to his retirement in 1990. To judge from *The Common Fisheries Policy: Origins, Evaluation and Future* (1994) Mike Holden is without doubt one of the Policy's sternest critics. A glance at the cryptic chapter headings suggests a mass of unresolved conflicts over policy objectives and the choice of mechanisms that some six or seven years later remained virtually unchanged. Overall, after ten years Holden judged it to be a political success in the sense that the CFP had ridden out the potential storm of enlargement. Its priority task of keeping member states in harmony and onboard had been fulfilled. As a conservation policy, however, intent on controlling fishing effort and protecting the fish stocks it was an unalloyed failure, plagued by contradictions, ill-defined objectives and a poorly disguised unwillingness to take the problem of overfishing seriously.

Holden's critique revolved around a number of fundamental flaws in the policy design that rendered it un-operational. First and foremost was the failure to establish a clear hierarchy of objectives and to prioritise between the long-term conservation ambitions and the short-term benefits of improving the economic performance of the fishing industry. Arising from this was the incompatibility of the scientific and political interpretations of the Policy's intent: thus, the scientists were given no clear guidance as to the policy objectives, while the politicians were unable to appreciate the significance and urgency of the scientific advice and the consequences of non-compliance.

A further issue was the selection of inappropriate mechanisms for the articulation of the conservation goals. The case in point was the use of TACs as the flagship device for capping or reducing fishing effort. Total allowable catches are merely numbers describing the estimated safe annual catch measured in tonnes of fish that may be taken from a particular stock without the risk of damaging the stock's ability to regenerate at roughly the same or higher levels in future years. Within Europe's 'common pond', however, their primary function was to guide the allocation of shares of total catch among member states according to relative stability.

In relation to conservation, TACs are at best targets, requiring carefully planned deployment of a range of measures to ensure that they are met – including individual vessel quotas, by-catch restrictions, gear regulations, seasonal or permanent ground closures *inter alia*. In such combinations, TACs can work tolerably well in regulating fishing pressure in a single species but are ill-suited to the management of mixed fisheries. However, the efficacy of TACs was at the time further weakened by the unreliability of official landing statistics, uncorrected for the manipulation, wilful or otherwise, of actual catches. Unsatisfactory though they may be, when used in combination with a reliable quota management system, there is no better mechanism for both setting the targets and measuring the likely impacts on fishing activity. A final flaw in the design of the CFP was the member states' inexperience and lack of investment in systems of surveillance and enforcement. In these last decades of the 20th century, if a skipper was intent on thwarting the regulations the odds against them being apprehended were huge.

What is perhaps more surprising than the catalogue of errors is the apparent reluctance of the governing system to put things right. The generally preferred explanation is that the CFP was simply too fragile a structure to be radically altered without the risk of destabilising its acceptance by the member states and so leading to its possible collapse – to the humiliation of the Commission and Council. Thus far the CFP had survived, even if it had not succeeded in its principal task *vis-à-vis* the fisheries, thanks to the success of relative stability. But did relative stability really survive unscathed? We are told that it did – and for most member states this was probably true – but the facts may say otherwise in the case of certain other member states. Taking into account post-allocation swaps of fishing rights and the differential rates of stock depletion reflected in national TACs, in the UK's case in particular, relative stability is something of an illusion.

The reluctance, especially on the part of the Commission to intervene, also draws attention to its remarkable degree of inflexibility in choosing what course of action to follow and when, and its facility for apparently kicking difficult but essential issues into the long grass. All of this is indicative of a lack of critical oversight of its actions other than through the EU's own Court of Auditors – in other words, the apparent democratic deficit which the EU's critics are keen to assert.

While it is difficult to quibble with Holden's analysis, it is rather more difficult to accept his solution. Holden was very much against tinkering with a failing policy. Instead, he wanted to see a wholly different approach with a new structure, clearer objectives and the appropriate tools – one that respected the Treaty of Rome principles but shifted its focus away from the elusive goal of stock conservation and towards sound economic fleet management and the pursuit of maximum economic yield or MEY. He also advocated greater use of the 'subsidiarity principle' confirmed under the Maastricht Treaty (1992) that would allow operational decisions to be taken much closer to the point of delivery rather than at the centre. On this last point, I am in full agreement for subsidiarity is the enabler of both co-management and regionalising the CFP.

However, Holden pins the flag of MEY to the mast of the CFP rather too firmly for my liking. This much narrower economic perspective might perhaps coincidentally help to resolve the dilemmas surrounding the conservation of fish stocks but 'perhaps' is an insufficient basis for tackling the essential, long-term sustainability of the resource. As it turned out opting for MEY as opposed to MSY as the core objective would have put the CFP at odds with the United Nations' commitment under the World Summit on Sustainable Development 2002. The wiser course would be to stick with the aims and objectives of the CFP as it stood but to guide the Policy on a more clearly marked path towards sustainable fisheries. This would need to be signposted by much clearer statements of intent, greater attention to the practical operationalisation of policies, a lengthening of the timescales for management plans, greater flexibility in regard to the opportunities and limitations afforded by the different regional seas and stronger participation on the part of the fishing industry in both formulation and implementation of policy.

Looking ahead almost 20 years to the time when the UK fishing industry would be giving almost unanimous support to Brexit and withdrawal from the CFP, it is clear that there was little for the industry to celebrate in this first phase of the Policy's development. Most of their worst fears were being realised, notably in the steep decline in fishing opportunities in the offshore sector. For some it meant the growing threat of having to face decommissioning and premature retirement; for others, there was the likelihood of further tightening of the screw through 'days at sea' restrictions. The future outlook seemed bleak. The UK's two leading fisherman's associations (SFF and NFFO) were united in the quest for a return to 'home rule' in the shape of 'zonal management'. There was, however, little support for such a radical move elsewhere in Europe or among the UK's governing institutions.

Further reading

Holden, M. (1994). *The Common Fisheries Policy: Origin, Evaluation and Future.* Oxford: Blackwell Scientific Publishing.

Lado, E. P. (2016). *The Common Fisheries Policy: The Quest for Sustainability.* Chichester: John Wiley & Sons Ltd.

Symes, D., Crean, K., & Phillipson, J. (1996). *Devolved and Regional Management Systems for Fisheries. Final Report.* Hull: University of Hull.

See also references cited in Chapter 4 in relation to co-management and regionalisation.

6

ALTERING COURSE

2003–2012

Introduction

Something had to give. It was no longer possible to continue with a failing policy that had already presided over an unparalleled 20 years of decline for Europe's fishing industries. There was, however, little appetite for ditching the entire project. The UK industries were keen to see a system of decentralised 'zonal management' for each of the major sea areas that make up the Atlantic component of the common pond replace the highly centralised CFP. But this was an idea well ahead of its time and lacking in detailed structure. By the end of the 20th century, the CFP was seemingly firmly anchored and its resilience well tested, but it was not yet functioning properly. The task of the 2003 relaunch was therefore to reaffirm the CFP's relevance in a rapidly changing world, to prune some of its dysfunctional features and above all to improve its implementation. One thing was already quite clear: nothing happens very quickly with the CFP and radical change is all but unthinkable.[1]

Blocking the path to any radical transformation of the CFP were three key factors. The first was path dependency with relative stability – the principal weapon against transformational change embraced by almost all member states. The second was the emergence of a distinctive north–south divide between the Friends of the Fish and the Friends of Fishing with the latter capable of forming a blocking minority should it come to a vote. The third and possibly the most stubborn obstacle was institutional inertia – the Treaty of Rome's apparent iron grip on the basic structure of the EU decision-making machinery: there was simply no room within this governing system for an intermediate level of decision-making that might be seen to challenge the authority of either the EU's governing institutions or the member state.

DOI: 10.4324/9781003362913-8

However, despite the continuing conflicts of interest among member states and the apparent discord over different approaches to improving the functioning of the CFP, there was an increasing awareness on the part of member state governments and the fishing industries of the onset of a state of crisis concerning the status of the basic resources and an emerging realisation that the industry must be prepared to face the harsh realities of survival and the need for adaptation. There was broad agreement that structural policy and particularly the role of subsidies and grant aid was aggravating the situation rather than addressing it. Cod recovery was fast becoming a hot political issue and there was a need for greater 'buy in' from industry stakeholders for both more effective regulation and to sustain the pressure over time without the frequent disruption from politically motivated annual decision-making. There was also the stirring of the marine environmental lobby – a potentially dangerous enemy for the fishing industry, capable of mobilising considerable popular support. Thus the system to be governed appeared to be in a better state of preparedness for more effective regulation.

There was, however, surprisingly little scope for action. Rule out radical change and banish any attempt to interfere with the basic design of conservation policy and one is left with only one area for intervention: improving the efficiency and effectiveness of the implementation process by broadening the expert advice, increasing the awareness of regional diversity and focusing attention on the long-term sustainability of the fish stocks and their ecosystems as the basis for securing the future sustainability of the fishing sector.

2003–2012: trimming the sails

The prospects for a revised CFP were presented in a Green Paper that gave a pretty frank account of the Policy's earlier failings and set out the reality of the situation facing Europe's fisheries. It provided a basis for consultation with member state administrations and with fishers and other stakeholders, setting minds at rest that the treasured relative stability was in no way compromised by the proposed programme of action. There were, in Penas Lado's words, three headline objectives: the ending of funding for new vessel construction; the establishment of Regional Advisory Councils; and the creation of long-term plans for fisheries management on a single species basis. Other concerns, expressed through separate communications covered closer environmental integration, the need to address the discards problem and the development of aquaculture. But there was no specific mention of the issue that dogged fisheries management in the EU's northern waters of how to manage the prevailing mixed fisheries.

Ending financial support for new vessel construction

The new basic regulation (2371/2002) contained, as expected, a major hammer blow for the fishing industry – the removal of the hitherto generous financial support for new vessel construction, previously the industry's principal route to a modern, efficient and competitive fishing fleet. That provision had run counter to and in fact undermined the main objective of the conservation policy of reducing fishing capacity and thereby fishing pressure on declining stocks. Without strong action, the EU's fishing industry would be bound to follow a downward trend of lower catches, reduced profit margins and an increasing risk of bankruptcy.

Its introduction was inevitable and despite considerable opposition from the industry the only concession was a brief phasing in of its implementation with a delay of two years for vessels under 400 GRT. As Penas Lado points out, the Commission's decision was influenced by external as well as internal policy concerns. The World Trade Organization (WTO) was troubled by the impact of publicly financed grants and subsidies on the principle of 'fair competition' within a global trading environment. The Commission's concern was largely over the growth of fishing capacity. It was clearly not feasible to ban or cap new constructions for that would lead to an ageing, less efficient, less viable fishing sector. The ending of grant aid did not in fact bring new constructions to a halt; it merely shifted the onus for funding the modernisation of the fleet from public to private sources, increasing the burden of indebtedness. In doing so it also brought the harvesting sector more firmly within the realm of private capital, incidentally increasing pressure for further market initiatives for the privatisation of fishing rights.

However, the ending of direct financial support for new vessel construction did not bring closure to the issue of capacity expansion. A host of other financial measures aimed at increasing the operational efficiency of the fishing vessel, reducing costs and improving quayside installations, markets and processing facilities indirectly contribute to the enhancement of fishing capacity *per se*. Whereas fleet capacity might appear to be reduced in terms of vessel numbers and gross registered tonnage, the technological efficiency of the fleet continued to outstrip the decline in vessel numbers. Only very gradually was the portfolio of financial support to shift its emphasis from capacity-building measures to what may be described as 'beneficial subsidies' that directly improve sustainability of the fish stocks.

Regional Advisory Councils

The introduction of RACs was the least controversial, most widely and warmly welcomed initiative. Welcomed not only because at last it gave the industry a direct line of communication to the governing system's decision-making elite and an early opportunity to familiarise itself with new policy proposals, but also because it provided policy makers and managers with access to the vast experiential knowledge of the fishers and specific regional intelligence that might assist the design of new policy initiatives to reflect and fit the needs of particular regional fisheries.

Regulation 2371/2002 provided for the establishment of five Advisory Councils with a regional designation and two dealing with pelagics and distant water fisheries. Their remit was to advise both the Commission and relevant member states of industry-based views on Commission proposals and self-generated 'own account' concerns. There was, in the early years, no guarantee as to how – or even if – the RACs' advice was taken into account and no formal feedback from the Commission. RACs were clearly intended to perform a consultative rather than negotiative role. The RACs were rolled out over a period of five years with the North Sea in 2004 quickly followed by the North West Waters, Baltic and South West Waters RACs in 2005, 2006 and 2007 respectively. The last of the complement – the Mediterranean – was established in 2008.

In their own right, and within the context of the CFP, RACs were in fact quite radical concepts. One of the more controversial elements was the balance of representation within the RAC: many in the fishing industry favoured exclusively fisheries-related membership. In the event, two thirds of Council seats were allocated to fishing interests and one third to a wide range of 'other interests' including the environmental NGOs reflecting the increasing relevance of the marine environmental matters in debates over fishing. Thus, RACs brought together representatives from the fishing industries of the relevant member states which hitherto had seen themselves as rivals and whose views on policy matters did not necessarily coalesce. Moreover, the inclusion of environmental NGOs, often seen as the 'enemies' rather than allies of the fishing industry, made for further tensions within the advisory process. Yet together they were expected to reach an agreement or, at worst, a compromise over key management issues before presenting their conclusions to the Commission. The broad membership of RACs sometimes appeared to exert 'a paradox of public participation' whereby the greater the range of public participation, the less influence can be influenced by the traditional actors – the fishers.

It would initially take time for each RAC to find its own *modus operandi* resolving what had previously been immovable objects to internal negotiations and also building a workable relationship with the Commission and finding the most effective way of communicating their ideas on a range of potentially difficult issues. RACs thus faced a steep learning curve but they were soon to become a crucial element in the complex process of reaching solutions to different policy concerns. Not all fishers associations were quick to applaud the RACs' performance; many were left with a sense that they were not being taken seriously enough in the corridors of power in Brussels and many in the industry remained highly critical of RACs' measured advice to the Commission.

Long-term fisheries plans

The idea of long-term management plans had first been included in the 1992 Regulation but never carried through into a formal proposal. Its current genesis lay in the cod crisis and a long-standing awareness of the need to ensure the long-term health of all heavily exploited stocks and guarantee their commercial as well as biological survival. Attention was initially focussed on a handful of commercial stocks compromised by overfishing but essential to the viability of Europe's fishing industry, thus leaving a large number of other stocks exposed to the risk of attracting greater fishing effort. The new approach required the definition of specific reference points based on a 'precautionary approach' to fisheries management.

Close observance of these reference points should, in theory, secure sustainable fisheries within sustainable ecosystems. The new approach was intended to cover both single-species fisheries and mixed fisheries though in practice current scientific assessments are only available on a single-species basis. Again, in theory, long-term plans should allow a more sensitive, adaptive form of management – an excellent vehicle for delivering an ecosystem-based approach though severely hampered by a lack of multi-species scientific assessments. Moreover, it was expected to provide fishers with greater stability on which to plan their own longer-term operations.

The new basic Regulation chose to base the objective for the multi-annual plans (MAPs) on the recovery or maintenance of stocks within 'safe biological limits'. It also recognised that the existing tools of annual TACs and catch quotas were insufficient for the task in hand. An additional constraint of effort controls (days at sea allocations) was initially considered a necessary element of the plan, much to the anger of the industry whose members now felt bound hand and foot by the regulatory process. Reflecting this concern the new policy was presented under two separate Articles – one for 'recovery plans' dealing with stocks currently outside safe biological limits that included effort controls and the second for 'management plans' for stocks within safe biological limits without the need for additional controls.

It was an open secret that the Commission had hoped to gain full control of the multi-annual management process and avoid the yearly exposure to political intervention from the December Council of Ministers whose task it was to fix the annual TACs. This was, not surprisingly, unacceptable to the member states and a compromise was reached whereby the Council would be involved only in 'exceptional circumstances' where the biomass was below the minimum level and the stocks were at high risk of further deterioration where the biomass was below the minimum level and the stocks at high risk of further deterioration. Sadly, from the Commission's standpoint such 'exceptional circumstances' were to prevail throughout the early years of the MAPs for certain key species.

The cod recovery plan

The cod recovery plan covering stocks in the North Sea (including the Kattegat and Skagerrak), West of Scotland and the Irish Sea was initiated in 2004. At the time the cod fishery was in crisis, its stocks were in real danger of commercial if not biological extinction with the spawning stock biomass (SSB) declining from circa 250,000t in the early 1970s to around 50,000t by the end of the century. ICES warnings were becoming ever more strident and its recommendations for zero TACs simply repetitive. The target set for the recovery plan was 150,000t and this was subsequently accepted as the basis for the maximum sustainable yield. It was a huge challenge and would require the setting of rigorous constraints on fishing activity in terms of both catch and fishing effort. The plan quickly took centre stage on the agenda of the newly formed North Sea RAC. Were the recovery plan to achieve its target it would be seen as a major step in the development of the CFP and proof that the Policy was working.

Three major issues confronted those responsible for managing the recovery plan. The initial question concerned its intended lifespan: should it be accomplished in a shorter timespan thus putting greater stress on the industry or over a longer period lessening the immediate pressure on those engaged in the fishery but prolonging the pain? The second and related issue concerned the so-called 'harvest control rules', the formula for calculating TACs that would observe a reasonable balance between the dominant conservation imperative and the economic and social concerns for the viability of fisheries enterprises reliant on the cod fishery. The final concern was how to manage the highly controversial 'days at sea' allocation given the diversity of vessel sizes and the *métiers* involved. At this point, it is perhaps useful to remind ourselves that cod is not a single species fishery but part of the very important mixed fishery that lay at the heart of the region's demersal fishing industry: the cod recovery plan thus had a potentially wide impact.

Initially, the cod recovery plan was set as a ten-year programme with the TACs set to generate a 30% annual increase in SSB. This was to prove an impossible task and, following the very weak signs of recovery in the early years, the plan was eventually converted into a rolling programme. Other single species MAPs followed quite quickly, with those for the northern and southern hake and Nephrops in 2005, with the Bay of Biscay sole in 2006 and the remainder of the initial tranche involving the Western Channel sole, North Sea sole and plaice and Baltic cod in 2007 – all dealing with stocks less perilously placed and each with its own strategy designed to match the particular conditions of the stock concerned.

Other issues

The 2002 reform was not solely concerned with these three items – the ending of grant aid for new vessel construction, the creation of RACs and the introduction of long-term management plans – though these were the essential elements of the revised conservation policy. There was considerable activity over improved technical conservation measures with particularly heated arguments relating to proposals for increasing the selectivity of fishing gears largely concerned with enlarging the mesh size and the introduction of square mesh panels in trawl nets to facilitate the escape of undersized fish. Improving the efficiency of control and enforcement measures was important in reducing illegal fishing activity within the common pond and ensuring that TACs were being properly observed. This latter matter was essentially the member states' responsibility, but a new control regulation was introduced in 2009 including an obligation for larger vessels to employ an 'electronic logbook' for reporting catch information and thereby providing a more accurate picture of quota uptake.

Possibly the more important concerns related to what was actually missing from the 2002 Regulation. Amazingly there was no concrete proposal for tackling discards and little apparent activity within the Commission apart from a report in 2006 on the state of affairs in EU waters. This indicated that in the North Sea, for example, on average 20% of the catch from trawls and seines was discarded rising in the case of beam trawling to discard rates of 40–60% of target and non-target species. Although the 2002 Regulation laid claim to the incremental adoption of an ecosystem-based approach to fisheries management, there was little evidence that this meant anything more than the protection of commercial fish stocks and the avoidance of damage to habitats and ecosystems. Indeed, most of the action to achieve these goals was being taken under the Marine Strategy Framework Directive (2008) put together by the Environment Directorate.

The most obvious omission, however, concerns the goal of maximum sustainable yield (MSY) – not so much a slight or oversight on the part of the Commission but an accident of timing. In 2002, at the conclusion of the UN's World Summit on Sustainable Development eight measures were agreed for achieving sustainable fisheries, of which restoring stocks to MSY by 2015 was the key target. Although a well-established concept among fisheries scientists (though not universally approved), its implications were not fully understood in management circles. There was too little time for a meaningful discussion during the closing stages of the CFP reform process and it was, therefore, not included in the text of the Regulation. It was, nonetheless, implicitly adopted in the minds of most policy makers and it became an increasingly significant reference point for the progress of the CFP.

Assessment

Trying to assess how well the revised CFP had performed in the first ten years is difficult and can at best only be regarded as an interim process. The two key improvements – RACs and MAPs – were only just up and running by the mid-point of the period and barely into their stride by 2012. The task is further complicated by the question as to how one judges 'success'. In terms of the conservation objective, there are at least three different though complementary measures: reduction in fishing effort; improvement in fishing mortality; and progress towards MSY. In addition, there is an even more complex question: how do you attribute credit or blame for what has happened? And let us not forget the economic and social objectives of the CFP.

By 2012, things were looking brighter. Certainly, the mood had changed from one of despair at the start of the 21st century with a clearly failing CFP to one of cautious optimism by 2013, the start of the second phase of revision for the CFP. There was a sense that maybe a corner was being turned and the Policy was beginning to make sense. Reductions in fishing effort were apparently being achieved but by what means? The size of the fleet was being cut in terms of the vessel numbers, tonnage and engine capacity but the estimated aggregate technical capacity of the EU fleet remained roughly at the levels of the final decade of the 20th century. The implication was that the expanded fishing effort was declining as a result of reduced fishing opportunities (TACs and days at sea) but, despite the introduction in some member states of voluntary vessel decommissioning schemes, a greater share of the actual fishing capacity was simply being temporarily unused. By the end of the ten-year period, fishing mortality in many of the key commercial stocks including cod was falling, but was this attributable to policy intervention or to natural causes? There was, for example, evidence of much-improved cod recruitment in the second half of the 2000s contributing to an uplift in SSB in the early 2010s that would account for at least part of the falling fishing mortality.

Judgement concerning progress towards MSY status was made more uncertain partly because MSY had not been formally adopted as a policy objective in the 2002 reform. However, the situation again looked quite promising. In 2005 94% of the 34 EU fish stocks under MSY assessment were being overexploited; progress in the next four years remained weak but by 2012 the proportion of the 38 assessed stocks that were overexploited had fallen to 47%. Just how secure this apparent progress towards sustainability of the EU's fisheries was, will be considered in the next chapter.

Given these tentative signs of achievement with regard to the conservation objective, how well did the economic and social objectives of the CFP fare in this very challenging period? On these issues Penas Lado is disappointingly silent; the evidence in fact allows for much less positivity. Reduced fishing opportunities meant smaller catches for the majority of fishing enterprises, lower revenues and weaker profit margins. With a growing level of unused capacity and increasing direct costs, the fishing industry in general was becoming less economically efficient. The EU's own data on financial performance indicated that in the mid-2000s net profits totalled a mere 6.4% of landed value – a level well below that required for healthy economic activity. Six out of 13 reporting member states reported net losses. The EU's fisheries were becoming economically stressed.

Taking a closer look at the performance of the Scottish fishing industry in 2008 (Scottish Government 2010), it becomes clear that average performance indicators for the fishing industry as a whole conceal significant disparities between (and within) the different sectors. Despite undergoing downsizing of the fleet in the late 1990s and early 2000s mainly as a result of decommissioning schemes, the demersal sector had edged even closer to financial disaster and was now on the knife edge of viability. In 2008 average net profits per trawler were a very modest £8,400 with most vessels in the lower quartile returning net losses. Returns were somewhat stronger for the quota-regulated Nephrops trawl fishery (£16,800). But this fishery which had earlier shown strong growth and enjoyed high prices on the European market was increasingly being seen as vulnerable to increased fishing effort, low investment and weakening market prices. Scotland's pelagic sector, however, was enjoying a boom period with stocks and TACs at comparatively high levels, especially for mackerel, a much-slimmed down fleet and buoyant markets. The pelagic industry had benefitted significantly from strong investment in developing a modern, technically very efficient fleet of large mid-water trawlers. Like its European counterparts, the Scottish pelagic industry came closest to the model of a well-organised, efficient and disciplined sector – regulated by international agreements spanning the NE Atlantic articulated through the North East Atlantic Fisheries Commission.

Social issues were to a large extent lost, or at the very least forgotten, causes in the realm of fisheries policy and management, where policy makers usually narrow the definition of social issues to cover employment, working conditions and wages. Employment in the EU's fishing industry had continued to fall steadily throughout the lifetime of the CFP. Reduced on-board employment was a necessary corollary of a modernising fishing industry where improved technical efficiency almost invariably involves the substitution of capital for labour. The fall in employment was further accelerated by attempts to remove excess capacity. Rising operating costs were forcing vessel owners to cut the size of crews; even within the family operated small scale enterprises it was becoming the norm to reduce the numbers on board from three or two to one, though in this case the cause may have more to do with a lack of available labour.

At the start of the 21st century, the concern over employment had less to do with declining job opportunities and more to do with the issues of recruitment. In addition to outflows of young people from remote coastal regions in search of the 'bright lights' of urban living, there was a growing lack of confidence in fishing as a source of reliable, well-remunerated employment. Moreover, the ambitions of those wishing to build a career within fishing were blunted by two factors: the financial costs of becoming a skipper-owner in terms of buying a second-hand vessel, fishing gear and fishing entitlements; and the severely damaged image of the skipper-owner as a self-reliant, independent, individual well able to navigate their own course to success.

There was concern also over the protection of individual fishing rights. The informal adoption of market based 'solutions' in some member states to the issue of balancing limited available resources to existing fishing capacity through individual transferable quotas was beginning to threaten the opportunity for future generations to gain entry to the industry and so put at risk the very survival of small fishing communities.

Note

1 When giving oral evidence to the House of Lords' European Committee's inquiry into the CFP, I was asked how long it might take for a new idea like regionalisation to become part of CFP thinking: my instinctive response was ten to fifteen years.

I initially took their Lordships' collective intake of breath as an intimation of my pessimism. Now, on reflection, I am tempted to think of it as a rebuke for my rashness. In the case of regionalisation, it is likely to prove a gross understatement.

Further reading

Lado, E. P. (2016). *The Common Fisheries Policy: The Quest for Sustainability.* Chichester: John Wiley & Sons Ltd.

See also:

Gray, T. S. (Ed.). (2005). *Participation in Fisheries Governance,* Vol. 4. Dordrecht: Springer Science & Business Media.

Symes, D. (2005). Regionalisation of Fisheries Governance: An Empty Vessel or a Cornucopia of Opportunity? in Gray, T.S. (eds.) *Participation in Fisheries Governance. Reviews: Methods and Technologies in Fish Biology and Fisheries,* Vol. 4, pp. 85–102. Dordrecht: Springer.

7

A RADICAL AGENDA MEETS INSTITUTIONAL INERTIA

2013–2020

2013

The mood going into the review period preparatory to the 2012 revision was uncertain, some might say unsettled. Progress with the long-term management plans had initially been slow but was gathering pace, admittedly aided by what would prove to be short-run improvements in stock recruitment for key species like cod. Herein lay a dilemma: was it to be simply a case of taking a fairly pragmatic stance, focusing attention on the recovery programme and redoubling the efforts of MAPs to deliver MSY? Such was the view endorsed by most fisheries scientists and some within the governing system. Or was it now time for a more ambitious, radical rethink of EU fisheries management that would bring other, so far neglected management approaches into play and hopefully reenergise policy makers, managers and the fishing industry and give the latter more responsibility for reshaping their future? In essence, this would mean reassessing and reforming a governance model for EU fisheries that was judged to be falling well short of the norms of good governance – a view considered timely by most economic and social scientists as well as many in the industry (though each for very different reasons) and by some within DG Mare itself. Thus, as the decennial review and reform process was taking shape, the CFP was at a crossroads: would its future direction be determined by a radical agenda for change or by instructions to maintain the existing course?

The issue of governance was a burgeoning theme in academic discourse throughout the late 1990s and early 2000s (see Kooiman et al., 1999) and later developed by Kooiman et al. (2005) into a model that sought to integrate the 'system to be governed' and the governing system into a functioning whole through the strength of interactions between the two sub-systems. Usage of the term 'governance' tends to vary in meaning: seen from within the governing

DOI: 10.4324/9781003362913-9

system it is interpreted more as a process that can be manipulated to give better policy outcomes, but outwith the system by industry, for example – it is more of a mechanism, generally seen as unyielding and insensitive to the realities of fishing. For the academic, it is essentially a theoretical construct, flexible in design and practice, and adapted to the prevailing political culture.

As a governing system, it is clear that the CFP was suffering a serious image problem. In their report to DG Mare (see below), Sissenwine and Symes (2007) suggest that the Commission was seen by the industry as a 'regulator and enforcer' rather than as a 'facilitator or enabler' – remote, unresponsive and bureaucratic in its relations with the industry and thus losing the confidence of its client group and the public at large. It is not too difficult to identify two main causes of this damning indictment: the narrowness of the policy-making process and the intransigent 'command and control' means of delivering policy decisions. The narrowness of the policy formulating process in respect of conservation was unique among the EU's sectoral policy domains largely in respect of the exclusive competence granted to the Commission. This exclusivity was further enforced by the role of the Council of Ministers as the sole arbiter of decision-making acting, as it had done, without the consent of the European Parliament; this was due to be amended following the signing of the Treaty of Lisbon (2009).

There were, perhaps, some sound reasons behind this seemingly undemocratic state of affairs. It was considered necessary to enable the imposition of a robust, coherent and consistent approach to stock management within the common pond that only a single, autonomous authority could command. Negotiated solutions (still less, independent member state solutions) were likely at best to result in weak compromises and a lack of consistency and at worst in internally contradictory and even more dysfunctional management systems. But there was clearly an urgent need for greater accessibility, transparency and proportionality in handling key issues and for much greater synergy between the EU institutions and member states, regional and local organisations and private enterprises in order to make policies work more effectively.

However, there was a further crucial dimension to the governance question. By 2012 the 'common pond' occupied a very large area extending from the Gulf of Bothnia in the north to the Canary Islands off the coast of northwest Africa spanning 40 degrees of latitude, and from the Azores in the west to Cyprus in the east – a range of 60 degrees of longitude. The fisheries within this huge, irregularly shaped maritime space are highly diverse yet subject in effect to a single, largely monolithic CFP.[1] The combination of such a large, diverse 'system to be governed' and the highly centralised, inflexible governing system only served to heighten the sense of remoteness, suspicion and even antipathy felt by the fishing industry resulting in low levels of commitment to and unwilling compliance with the regulatory framework. It was an unlikely basis for effective, interactive governance.

Thus far the EU institutions had made very few concessions towards bridging the yawning gap between the governors and the governed. Stakeholder consultation was limited. An advisory committee had originally been set up as early as 1971 to assist in matters relating to structural and marketing aid and a much-revised version – the Advisory Committee on Fisheries and Aquaculture – was created in 1999 with four separate working groups (fisheries, aquaculture, markets and structures) to both advise and lobby the Commission. Its influence appears to have been muted and later its role was partly subsumed by RACs. Establishing the RACs in 2004 represents the only substantive move to create an active role for the fishing industry within the formal policy process. Otherwise, contact with the industry was informal, infrequent and haphazard. Was it possible that the imminent review and reform process would finally provide for a much closer, participatory engagement with those who fished for a living?

The review and reform process

The review process was far better prepared than on the two previous occasions: the scope was much broader, the Commission's own discourse more stimulating and the consultation with interested parties more thorough. It was formally launched with the publication of the Commission's Green Paper in 2009. But in reality it had begun somewhat earlier. In May 2007, together with Mike Sissenwine, a recently retired leading American fisheries scientist, I was invited 'to reflect on the CFP and fisheries management in Europe': we were given no more detailed a remit and the time allowed for our reflections was short, leaving little time for collecting new evidence. We were anxious, however, to tap into the experience, opinions and aspirations of those directly involved in the formulation of EU fisheries policy, including key members of European and other stakeholder interests, and managed to cram in a significant number of informal discussions with a wide range of interested parties, sufficient to broaden our own perspectives.

When it came to writing the report, we opted for two separate but complementary commentaries – one analysing where the conservation policy had gone wrong and how it might be corrected; the other reflecting on institutional issues and the means of creating greater synergy between the CFP and the fishing industry. What emerged was a fairly hard-hitting critique of the CFP both as an agency for fisheries conservation and as an instrument of good governance, intended to stimulate further dialogue within DG Mare rather than prescribe solutions to specific problems.[2] In the following year, an internal paper was produced by DG Mare in which a number of the issues raised were more fully explored.

When the Green Paper was published in April 2009 it contained the outlines of a remarkably bold, innovative and radical agenda for reforming the CFP – albeit its style was softer and the proposals rather more implicit than explicit. The contrast between its vision of a medium-term future (2020) and the 'current reality of overfishing, fleet overcapacity, heavy subsidies, low economic resilience and decline in the volume of fish caught…' was stark. The CFP needed, in the Commission's words, 'a whole-scale and fundamental reform', not 'yet another piecemeal incremental reform'. Its vision was for MSY as a firm obligation, with fishing activities based on economically sound principles and fleets capable of adapting to both environmental and market change.

As a basis for reform, the Green Paper uncovered five structural failings to be used in identifying key areas of radical action:

i) 'fleet overcapacity' that could be addressed through the introduction of rights-based management to promote greater efficiency and create a better balance between fishing capacity and the resources;

ii) 'imprecise policy objectives' that could be tackled through the prioritisation of ecological/ stock sustainability;

iii) the Policy's 'short term focus' that disrupted the continuity of policy measures and weakened the flexibility of the industry's fishing strategies that could be improved by long-term regional management plans implemented through member states acting collaboratively within clearly defined limits set by the EU institutions resulting in simpler regulation sensitive to local conditions and designed with the help of industry participation;

iv) 'insufficient responsibility for the fishing industry' could also be remedied by the introduction of results-based management (and a strengthening of the role of FPOs) that would combine the industry's rights, responsibilities and accountability; and

v) 'lack of political will to ensure compliance' to be dealt with through unspecified reforms of the control and enforcement protocols.

To these five strategies for redeeming existing structural failings was added a range of 'further improvements' including a differential approach to managing small-scale fisheries; making the most of our fisheries through the implementation of MSY; a broader 'from catch to consumer' perspective for fisheries management; the integration of fisheries within a more comprehensive integrated maritime policy arrangement; and an overhaul of relative stability *inter alia*. With each of the issues raised (structural failings and further improvements) the Green Paper included a short list of questions around which interested parties could structure their response.

With almost 400 written responses and around 150 separate 'face to face' discussions mainly with member state and stakeholder organisations the consultation was both extensive and comprehensive. The responses overall were positive and suggested that there was a general inclination for a change from the routine path dependant changes of previous reforms. Both member state administrations and fisheries organisations were in large measure supportive of the direction of travel outlined by the Commission. Sadly, however, the thoroughness of the consultation process proved to be the reform agenda's undoing. Uncertainties and differences of opinion over the detailed implementation diminished the clarity of the overall response. It left the Commission with the all too familiar task of squaring the circle by trying to find workable compromises while maintaining the spirit of genuine reform.

The outcome: regionalising the CFP

For those who had pinned their hopes on a 'wholesale and fundamental reform' the outcome was disappointing. Some of the issues raised in the Green Paper had been dropped or 'relegated' to the level of member state responsibility, most notably the suggestion of introducing an EU wide system of rights-based management involving ITQs. Likewise, the proposal for a differentiated approach to the management of offshore and inshore fisheries with the former relying on ensuring its own economic viability through rights-based management, the latter dependant on fixed quota allocations and state-assisted viability. In both instances, there was more than a hint that the Commission was attempting to muscle in on what was traditionally member state territory. Nor were the member states prepared to cast off their protective clothing and sanction a restructuring of the relative stability principle.

In my own view regionalising the CFP was, and still is, the key to securing a truly radical reform of the CFP and failure to deliver a robust structure for achieving this is the most disappointing outcome of the 2012 review. This view is no doubt conditioned in part by my academic upbringing as a geographer in the 1950s that instilled a strong and lasting respect for the region as a meaningful concept in the organisation of functional space for human activity. More importantly, it seemed the most obvious way of attaining the common policy approach to cope with such a large and complex space as the EU's EEZ.

When the CFP was originally being negotiated in the 1970s it was required to fit the circumstances of the North Sea and the neighbouring northwest and southwest waters of the NE Atlantic: it was, in effect, a regional policy. As the enlargements of the EEC and EU unfolded in the 1980s, 1990s and early 2000s, this 'one size fits all' approach was stretched to breaking point. There was no longer any attempt, conscious or otherwise, to justify the Policy in terms of its relevance to the particular conditions and needs of the area's fisheries. What had once been a tailor-made policy had regressed into an ill-fitting generic approach.

To regain its relevance in the eyes of the fishing industry the CFP needed to be restructured around strategies designed to address the specific conditions and needs of each of the five regional seas – the Baltic and North Seas, North West and South West Waters and the Mediterranean. These separate regional strategies would need to be fully compliant with a common set of principles, norms and objectives established by the EU institutions and to draw upon a toolbox of instruments also sanctioned by the EU for use in implementing the strategy. To enhance the buy-in and willing compliance of the fishing industry, it would be essential to call upon and be seen to be implementing stakeholder advice available through the medium of the existing RACs. Direct involvement of the member states responsible for implementing the strategy would also be needed.

It might just be possible to achieve all this through the existing governing system but to secure the all-important sense of the industry's co-ownership of the regional project it would be preferable to devolve responsibility to independent regional agencies. Therein, apparently, lay the seeds of the project's self-destruction. In no sense would it mean creating fully autonomous regional authorities. Regulatory proposals for regional management would be subject to the oversight, scrutiny and approval by the relevant EU governing institutions before passing into law.

There were clear advantages to be gained from such a regional approach over and above a closer relationship between the governors and the governed and hopefully more appropriate management decisions. It would enable the removal of the burden of micro-management from the shoulders of the Commission. Opportunities for participative governance and co-management ventures would be greatly enhanced. The regional seas approach could also prove to be a vehicle for delivering other key aspects identified in the Green Paper. For example, it could more naturally facilitate the development of EBAFM and provide for closer integration of the CFP with the EU's marine environmental policy also structured along similar regional seas lines and its integrated maritime development policies.

So why was the regionalisation project, endorsed by DG Mare and widely advocated by the then Commissioner for Fisheries in several public pronouncements, finally abandoned to be replaced by a much less ambitious plan for neighbourly collaboration? The reasons appear unrelated to the viability of the project itself but rather to the risks that regionally devolved agencies might at some future point pose a threat to the authority of existing governing institutions. The principle laid down in the Treaty of Rome (1957) establishing the European Economic Community made no provision for any decision-making body other than the European Commission, Council of Ministers and Court of Justice and the member states.

There were concerns expressed that the creation of regional agencies might be seen as a first step towards nationalising fisheries policy (!) or that the agencies might at some future date challenge the authority of the EU's governing authorities. Possibly member states felt threatened but there was little or no indication of this in their responses to the Green Paper. With reasonable justification, some of the smaller states voiced concern that their fishing interests might be overlooked by the new agencies, though continued oversight of proceedings by the Commission would be a sufficient safeguard against such neglect. However, for reasons of institutional inertia or, worse still, a determination to protect the EU's institutions own competencies, the opportunity for fundamental reform of the CFP was let slip.

Despite the rejection of independent regional agencies as the major driving force for the future development of the CFP, all was not quite lost in the quest for a more regionalised system of fisheries management. The new approach, outlined in Article 18, is in essence a voluntary and delegated opportunity for neighbouring member states to collaborate in seeking solutions to shared problems of management within the same regional sea, subject to the Commission's approval. So far progress has been limited and mixed. Groupings of member states in the Baltic (Baltfish) and North Sea (the Scheveningen Group) have begun operating, taking advice from the relevant (R)AC but so far restricting their activity to relatively minor, local technical issues. Their influence to-date has therefore been muted and has not so far led to any regionally specific management solutions as originally hoped for. As Troels Hegland and Jesper Nielsen (2020) suggest the current move is simply a step away from the prevailing 'one size fits all' approach, but hopefully the first in a gradual stepwise progression towards regional management.

The other initiatives

But what of the other major initiatives outlined in the Green Paper? Perhaps not surprisingly those that made progress as the result of the 2012 review relate directly to the further development of the existing conservation policy. The first was the reaffirmation of the key role for science-led, long-term management plans covering the most commercial species and now firmly based on the objective of achieving MSY status by 2020 with Fmsy as the principal mechanism. Progress in the northern and western waters of the common pond has been encouraging with most stocks now exploited at MSY levels, in contrast to the Mediterranean and Black Sea where stocks continue to be fished at levels above those required for sustainability. There is, however, the precautionary tale of the enigmatic cod stocks.

When we last visited the cod recovery plan in the late 2000s and early 2010s, the cod stocks appeared to be recovering well, aided by a run of good recruitment years. By 2017 the spawning stock had achieved its target of 150,000 t, reaching MSY status and was receiving the Marine Stewardship Council's (MSC) certificate as a sustainable fishery. Only two years later the situation was reversed. According to ICES, the spawning stock had fallen well below the safe biological limit of 150,000 t – and therefore well outside the MSY target level – and its new advice was for a staggering 64% cut in the TAC for 2020.

The final decision to opt for a more modest 50% reduction still left North Sea fishing interests fearing 'the end of the fishery as we know it'. The combination of a very low TAC and discard ban with cod stocks still well above the historic minima of the early years of the decade was deemed likely to trigger a very early closure of the vital mixed demersal fishery while quotas for haddock, saithe and whiting were still available.

A 'hammer blow' for the EU and Norwegian demersal trawling industries, it also raises some serious questions over both science and management policy: just how secure is the science underpinning estimates of spawning stock biomass? and has the raising of TACs for the recently 'recovered' cod been a touch premature? With hindsight it seems wiser to assume that the recovery of seriously overfished stocks like the cod is more likely to be unstable and a more precautionary approach to its exploitation is advisable.

By far the most significant move was the long-awaited action to tackle the now chronic problem of discards that was making a mockery of the EU's attempts to achieve sustainable fisheries. The industry, while recognising the scale and urgency of the situation, was in general opposed to a discards ban, preferring instead to reduce the incidence and volume of discards through fine-tuning and wider application of technical conservation measures. Avoidance of the problem in the first place would be preferable to drastic action to solve it. However, such measures had in the past made little impact; the situation was not improving, and the Commission was left with no other option than a ban.

Implementing the 'landings obligation' – the Commission's weasel words used in Article 15 to describe the ban – was to prove extremely difficult and its task made that much harder by the previous failure to address the broader issue of managing the mixed fisheries. Here, the presence of 'choke species' – those species that because of their lower abundance within a mixed fishery and therefore much smaller TACs – was adding to the difficulties in handling the discard issue. Their presence threatens the premature closure of the entire mixed fishery when the quota for the choke species runs out, causing significant levels of unfished quota for other species with loss of fishing opportunities, reserves and profit.

Scheduled for implementation through a rolling programme over several years, the landings obligation has proved deeply unpopular with large swathes of the fishing industry, with criticism directed at its 'accelerated timetable', its perceived unworkability, fears of discrimination between different types of fishing, the threat to the viability of some fishing enterprises but, above all, because discarding is itself largely the product of the CFP's own mishandling of fisheries management. In the final year of the rolling programme, the situation was not fully resolved. While the Commission reported that the landings obligation was working well and stocks were improving, the levels of legally landed 'discards' were said to be low and there was growing scepticism among some in the industry as to whether it was being fully implemented. It will probably take several years for the new regulation to be bedded in, the necessary adjustments by both industry and management to be made and for a growing realisation to emerge that the landings obligation was making a significant contribution to the future sustainability of the fisheries.

Other initiatives featured in the Green Paper have made little progress. Once again, the notion of EBAFM was given scant attention. Its chances of further promotion probably died when a full-on approach to regional management was taken off the table. DG Mare's somewhat lukewarm endorsement in the Green Paper of the need for closer integration with marine environmental management seemed to shuffle off its responsibility for an ecosystem approach and place the onus on DG Environment's Marine Strategy Framework Directive and its detailed plans for delivering 'good environmental status' for the EU's seas by 2020. Of the Directive's four targets that relate to fishing, only one concerning sustainable fish stocks is currently integral to the CFP; others dealing with biological diversity, marine food chains and seabed integrity – all likely to be incorporated within a genuine ecosystem-based approach to fisheries – are effectively ignored.

As expected, DG Mare's original suggestion of an EU-wide system of rights-based management, involving 'transferable fishing concessions' (more EU newspeak for ITQs) and tradable throughout the common pond, was firmly rejected. Its intention was to accelerate the process of rationalising the structure of European fishing fleets through the removal of surplus capacity and so tackle the persistent imbalance between harvesting capacity and the resource. However, it came dangerously close to the realisation of a single pan-EU fishing fleet that many fishers and some of their representative organisations, looking to demonise the CFP, saw as the long-term ambition of the Brussels elite. Moreover, it infringed the member states' assumption of responsibility for quota management. All that was left of the original idea in Article 21 is the encouragement to member states to establish their own national systems of transferable fishing concessions. As many coastal states regard ITQs as tantamount to the alienation and privatisation of a common property resource, there has been little rush to oblige.

In 2013, however, a potential breakthrough for EU fisheries policy in its broadest sense could have been signalled by the inclusion of aquaculture within the text of the new basic Regulation. Hitherto, aquaculture's somewhat offhand, initially indirect, relationship with DG Mare and its predecessors, was gradually changing to an increasing concern for its further expansion partly to make up for losses of production in the wild capture sector. Compared to the scale and growth rates of aquaculture in south and east Asia (and Norway!), the EU industry is both small scale and sluggish. In 2011 it accounted for around 1.2 m tonnes valued at €3.5 bn, with production concentrated in four southern European states (France, Spain, Italy and Greece) that together with the UK contributed around 75% of both volume and value. Growth prospects are limited by a lack of access to available marine space and potential conflicts with other users, comparatively high levels of environmental protection, tight controls over the granting of new licences, a narrow range of species suited to 'caged' cultivation and the absence of 'big players' willing and able to expand (Penas Lado, 2016).

Although aquaculture has been included within the EU's grant aid schemes since the 1970s and subject to regulation in relation to marine environmental policy in the 1990s, it was only in 2013 that it was recognised as a major player in the EU's seafood sector and in the 'blue growth' vision for integrated maritime development and formally engaged with the CFP's mainstream objectives. As yet, a substantive regulatory framework for marine aquaculture production is missing.

Market and structural policies

An overview of Regulation 1380/2013 would suggest that the Commission has failed in its attempt to engineer a radical reform of the CFP and progress has largely been confined to strengthening the goal of sustainable fisheries. Even here, the only breakthrough has been with regard to discards legislation and the benefits of that have yet to be fully assessed. Much of the broader agenda concerning the future integration of fisheries policy within the EU's marine environmental and maritime development strategies appears to have stalled, perhaps only temporarily. Moreover, the tripartite structure for domestic fisheries policy founded on the integration of conservation, economic and social objectives seem to have been left behind.

To understand what has been happening *vis-à-vis* the economic and social objectives we need to look beyond the provisions of 1380/2013 that we all too easily think of as containing the body of CFP law. We need to focus attention on the contents of the two other complementary Regulations: those governing the 'common organisation of the markets in fishing and aquaculture products' (Regulation 1379/2013) and 'the common provisions of the … European Maritime and Fisheries Fund' (Regulation 508/2014).

Indeed, it was the new market regulation that helped to salvage part of the Green Paper's proposals for the transfer of more responsibility for management onto the fishing industry itself. The new regulation empowered Producer Organisations to take responsibility for both production and market plans for their members' catches in order to maximise the efficiency of quayside markets – the crucial link between the catching and processing/consumer sections of the fish supply chain by minimising the risk of shortages or gluts in supplies. The move also reaffirmed the Commission's intentions to assist the industry to become more financially self-reliant and less dependent on government support.

The more difficult challenge lay in the social dimension of fisheries policy, including the well-being of both fishers and fishing communities. From what we have learned so far EU fisheries policy is more about the fish than the fishers. As one who has spent quite some time on the study of small-scale fisheries and the welfare of fishing communities, I know only too well the difficulties of defining, analysing and trying to resolve social issues associated with fishing. What is clear, as indicated briefly in Chapter 4, is that the social values and structures of fishing and fishing communities have undergone transformational change as a result of the modernisation and urbanisation of the fishing industry and there is an argument to be made that the EU has been neglectful of the roles played by social institutions and values that have thus far underpinned the industry's resilience and sustainability through the difficult years of the late 20th century (Symes and Phillipson, 2009).

The elusive social dimension, described by Penas Lado (2016) as the 'least visible' of all the CFP's basic components – but nonetheless a perpetual thorn in the Policy's flesh – is, in fact, hidden among the contents of the so-called 'structural policy', itself an alias for the funding mechanism for supporting the fishing industry. This structural policy is the oldest element of the CFP and has undergone a number of different guises. It began in the 1970s as part of the European Agricultural Guidance and Guarantee Fund (EAGGF) before becoming the dedicated Financial Instrument for Fisheries Guidance (FIFG) for the period 1984–1999 and principally concerned with improving capacity and efficiency through generous support for new vessel building and modernisation of the fleet.

FIFG II (2000–2006) however coined the term 'socioeconomic measures' to fund early retirement schemes, support for small-scale fishing enterprises and technical innovation.

A real breakthrough for the social dimension, however, came in 2007 with the creation of a European Fisheries Fund (EFF) that embraced a comprehensive and more clearly defined set of objectives. The fund was divided into four so-called Axes: Axis 1 dealing with adaptation of the fleet; Axis 2 with aquaculture; Axis 3 covering infrastructure and environmental projects; and finally, Axis 4 concerned with the sustainable development of fisheries-dependent areas. The latest twist in the funding saga came in 2013 with the inevitable broadening of the remit to reflect the new emphasis on maritime development (Blue Growth) and another renaming as the European Maritime and Fisheries Fund (EMFF).

Axis 4 had come closest to the idea of a social dimension for fisheries policy. Its objectives were still couched in broadly socioeconomic terms with an emphasis on employment, including the economic diversification of fisheries and support for community-led plans and projects and strengthening the roles and capabilities of women within fishing communities. Its principal instrument was Fisheries Local Action Groups or FLAGs, modelled on the successful LEADER initiative for rural areas, which has secured some remarkable achievements, by adding value to local catches, creating additional employment, integrating the local small-scale fisheries sector within the wider local economy and above all creating a stronger basis for optimism in the future of the local fisheries sector. But a full assessment of Axis 4 and the role of FLAGs is still awaited. In the meantime, the history of instability in the remit of structural policy and a tendency to clutch at new fashions in the past tends to make one a little apprehensive over Axis 4's future and that of the FLAG initiative in particular.

As to the broader question of whether these new funding arrangements are sufficient proof of a social dimension within the CFP, the answer remains elusive. In terms of the CFP itself, the European Fisheries Fund (EFF) and the Axis 4 objective were certainly a marked improvement. Whether it is enough to nourish the resilience of fishing-dependent households, communities and areas and secure their social sustainability, the answer can only lie in the future.

The CFP 1983–2020: success or failure?

We have taken the critical review of the CFP just about as far as it will go. But we have one more task to fulfil before we close this unfinished story. We need to collect our thoughts and answer the final question as to whether the CFP has redeemed its earlier failures in taking effective control of managing some of the world's most important fisheries and set itself on course to tackle the difficult scenarios that lie ahead, notably around the impacts of global climate change. It would be useful to reach a provisional verdict before venturing into the final phase of the story of the last 50 years – one that concerns a second, somewhat less dramatic, seismic shift in the political geography of Europe's fisheries initiated this time by Brexit.

Looking back over the relatively short history of the CFP I have an overall sense of an amazingly ambitious journey into the unknown that had rather too often taken the wrong turning but bravely struggled on. Only in the last few years has the aspirational horizon seemed within reach. Currently, however, the CFP stands accused on five main counts: the greatly reduced status of the most important commercial fish stocks; the persistence of latent fleet overcapacity; failure to develop the promised EBAFM; failure to achieve an effective, stable integration of the three pillars of the EU's domestic fishing policy; and the institutional failure to cultivate a system of good interactive governance between the governing system and the system to be governed. Reaching a verdict on these charges will very likely be frustrated by two factors. The first is the inexorable division between the basically optimistic Europhile and pessimistic Eurosceptic viewpoints, which I suspect has a much wider application than just the Brexit torn Brits. The second is the difficulty in deciding where exactly to place the blame – whether on the EU's CFP, the member states' administration or the fishing industry itself.

We need not to retrace our steps on all five charges: singling out three key issues should be sufficient. First let us consider the substantive evidence relating to the current state of the fish stocks, the core objective of conservation policy. Here the evidence points to a fairly persistent decline in stock biomass for key commercial species up to and beyond the turn of the century, gradually reducing in strength and eventually replaced by partial recovery and finally reaching what we hope will be a new equilibrium under the guise of MSY targets. Cautious optimism would suggest that the CFP has recovered some of the ground lost in the first two decades and may well be in a position to secure sustainable fisheries in the near future, though at levels well below those prevailing at the start of the CFP.

In reaching a verdict we will need, however, to bear in mind three supplementary questions. First, to what extent was Nature, in the shape of shifts within the marine environmental regimes responsible for both the persistence of the long downturn and the more recent short-run recovery of key fish stocks, such that the CFP's weak policy measures only served to exacerbate the former and assist the latter trend? Second, how far was the situation in the late 20th century aggravated by poor compliance on the part of the fishing industry either in the form of wilful IUU fishing activity or as a result of careless disregard for the rules? And finally, with the cautionary tale of the cod in mind, just how secure are the futures for some of the other recovered stocks?

The second charge concerns an area of neglect by the CFP in its failure to pursue the claim of adhering to the principles of EBAFM and the promise of closer integration with marine environmental planning. One can almost sense a state of standoff between DG Mare and DG Environment. The former still believes it is acting in accordance with an ecosystem-based approach by minimising damage to ecosystems and their habitats. It does so largely by respecting the provisions of

DG Environment's Natura 2000 programme for wildlife protection and seeking to rebuild stocks of overexploited commercial species. While all this does make a valid contribution, the ecosystem approach has potentially a much wider reach, achievable through the application of a fisheries science based on the realities of interactive relationships between species within the ecosystem rather than the much simpler extrapolation of single species management policy.

Pursuing this largely unexplored territory could potentially establish the basis not only for a more effective, less stressful mixed fisheries management but also for achieving the promised greater integration of fisheries and marine environmental management. There is, however, a serious complication to this improved approach: the willingness and capacity of the EU's scientific advisers to undertake such a major challenge. Thus far, fisheries scientists have intimated no great enthusiasm for exploring the viability and potentials of the ecosystem-based approach.

The final indictment is directed not against the CFP *per se* but against the system of EU fisheries governance as a whole. This is perhaps the most intransigent of all the issues surrounding the CFP and it concerns the apparent rigidity of the governing system imposed by the EU's founding Treaty of Rome. Earlier in the chapter, I raised the issue of 'good governance', citing the work of Kooiman et al. (2005) that describes a system of interactive governance in which there is a strong degree of reciprocity between the governing system and the system to be governed. When the governance of EU fisheries is closely examined it soon becomes obvious that there is little complementarity and reciprocity between the two subsystems. The system to be governed, framed by the EU's extensive EEZ, comprises highly diverse ecosystems, fisheries and cultures. By contrast, the governing system comes close to the archetypal highly centralised, inflexible command and control model in which policy proposals for fisheries conservation are framed by a remote, elite bureaucracy – the European Commission – with minimal scope for interaction with the system to be governed, except through the channels of the (R)ACs. Policy decisions agreed by the Council of Ministers and European Parliament are handed down to member states to enact, allowing almost no flexibility in the way they are applied to what are very clearly diverse situations.

The Commission turned down the opportunity to modify its own overcentralised, autocratic mode for formulating policy when it rejected DG Mare's proposals for a regionalised approach to fisheries management that incidentally embodies the ecosystem-based approach. It seems unlikely that DG Mare will risk reintroducing the regionalisation proposal in the near future. This inevitably imposes a severe limitation on the direction of future change for the CFP and for the development of a more sensitive, flexible and adaptive approach to governance that is desperately needed in order to maximise the benefits of an enlightened CFP.

It is tempting to clinch the argument by comparing the performances of the EU and other countries engaged in fishing the NE Atlantic. At first glance, the results appear quite convincing. Both Iceland and Norway have succeeded where the EU patently failed in terms of achieving a secure base for the future sustainability of its fish stocks, while maintaining levels of output and establishing a good rapport between governing systems and the governed. However, this exercise does not survive closer scrutiny; the comparison is as meaningless as that between chalk and cheese.

It is difficult to find much, if any, common ground for comparing Iceland and Norway on the one hand and the EU on the other in terms of their geographical, economic and political circumstances. Although disadvantaged by their comparatively remote locations, Iceland and Norway are more than adequately compensated by their uncomplicated, extensive and richly resourced EEZs, only slightly offset by the somewhat limited range of commercially important species available. Fisheries play a very much more significant role in their economies and, in Iceland's case in particular, in their export earnings – meaning that fisheries occupy a more elevated position in the pecking order for political attention and fisheries are rewarded with effective participative governance.

By contrast, many EU coastal states emerged as losers in the geopolitical re-shuffle in the second half of the 1970s, with a complex of unequal but generally small, 'incomplete' EEZs that were welded together to form the spatial frame-work of the EU's common fishing zone. Fisheries in the EU account for only a minute share of the overall economy in terms of employment and GDP. In the early years, the EU was far from being a united entity at ease with the content and style of the CFP. As a result, the European Commission was compelled to devote much of its energy trying to resolve the conflicting demands for equity among several member states competing for access to already overfished stocks, to the detriment of their future sustainability.

Before judgement is passed, it is only fair to consider wider character references on behalf of the CFP. Despite its difficulties, the CFP has in fact presided over the development of a much improved, modernised and profitable fishing industry, though just how much of this is in any way attributable to the actions of the CFP remains a moot point. What is more strikingly clear is that the internal common market for fish and fish products has helped to create an effective and reliable seafood sector for the EU's vast customer market, no doubt assisted by investment in appropriate quayside infrastructure, processing and distribution systems. And, finally, the simple fact that the provisional verdict reached by Mike Holden at the end of the CFP's first decade that it was a 'political success' still holds good. There has been little or no support for disbanding the CFP or for a major revamp of the current system. Indeed, there has been a willingness to put up with its inadequacies in return for the somewhat mythical security afforded by relative stability – even though the concept is fraying at the edges rather badly.

It may come as something of a surprise that, despite the inability to act in the best interests of regional and local fisheries, I still believe the CFP is Europe's fisheries best hope for the future. The EU is better off with the CFP than without it. Renationalising fisheries management will not solve the sector's present and future problems. But it could risk creating a medley of divergent scenarios for the management of key shared stocks and possibly provoke acts of intrusive non-cooperation between neighbouring states eventually leading to a breakdown in the necessary collaboration over their effective management. This may not be the most persuasive testimonial for the CFP after 35 years of modest achievement. There will be an opportunity to test this finding when we come to survey the possible future for European and especially UK fisheries post-Brexit.

Notes

1 The geographic scale of the task has to be qualified by the fact that what we have so far discussed in terms of fisheries policy applies in practice only to the Atlantic part of the CFP's domain. Although Mediterranean fisheries are also afflicted by overexploitation, for reasons linked to the region's physical and political geography – the narrow continental shelf, non-application of the standard EEZ principles and the spatially limited and fragmentary extent of EU fishing interests – their management under the General Fisheries Commission for the Mediterranean (GFCM) set up in 1949 by the FAO, has been limited very largely to technical conservation measures.
2 Although DG Mare had intended to release the report, an embargo was placed on its publication. Shortly thereafter, it was leaked, thus affording it even greater publicity!

Further reading

Commission of the European Communities. (2009). *Green Paper: Reform of the Common Fisheries Policy.* COM (2009) 163 final. CEC: Brussels.

Hegland, T. J., & Raakjær, J. (2020). The Common Fisheries Policy in *The Oxford Encyclopaedia of European Union Politics.* Oxford University Press. https://doi.org/10.1093/acrefore/9780190228637.013.1099

Kooiman, J., van Vliet, M., & Jentoft, S. (Eds.). (1999). *Creative Governance: Opportunities for Fisheries in Europe.* Aldershot: Ashgate.

Kooiman, J., Jentoft, S., Bavinck, M., & Pullin, R. (Eds.). (2005). *Fish for Life: Interactive Governance for Fisheries.* Amsterdam: Amsterdam University Press.

Penas Lado, E. (2016). *The Common Fisheries Policy: The Quest for Sustainability.* Chichester: John Wiley & Sons Ltd.

Scottish Government. (2010). *The Future of Fisheries Management in Scotland.* Report of an Independent Panel. Edinburgh: The Scottish Government.

Sissenwine, M., & Symes, D. (2007). *Reflections on the Common Fisheries Policy.* Report to the General Directorate for Fisheries and Maritime Affairs of the European Commission. Brussels.

Symes, D., & Phillipson, J. (2009). Whatever became of social objectives in fisheries policy? *Fisheries Research*, 95 (1), 1–5.

PART III

A seismic aftershock

Brexit

8

BREXIT AND ITS IMPLICATIONS FOR FISHERIES

Introduction

The UK's decision to leave the EU, signalled in a referendum in 2016 and completed by the signing of a withdrawal agreement in 2020, brings to an end some 47 years of membership of one of the world's largest and boldest political alliances founded on the basis of a single common market for goods and services. For 37 of these years, key areas of UK fisheries management have been determined by the EU's Common Fisheries Policy whose chequered career we have been following in the previous section.

Now, in the third and final section of this reflective study, we move out of our comfort zone of trying to explain what has happened to Europe's fisheries and their management in the recent past to the more difficult task of envisaging what might happen in the future. This means moving from a situation where we had an abundance of factual evidence to sift through in pursuit of the truth about the CFP to one where we have the minimum of hard evidence and, in essence, only our informed intelligence to guide us through an array of very different alternative futures. We are still some way off knowing all the basic parameters of the post-Brexit future of the UK's and the EU's fisheries sectors. These parameters were outlined in the final negotiations over the new relationship between an independent UK and the EU (see Chapter 9), though their impact may not be fully revealed for some years to come. Crucially, the post-Brexit future for UK fisheries will also depend on whether the political integrity of the United Kingdom actually survives the challenge of separation from the EU and its single, common market.

The final three chapters attempt to set out the basic conditions of a post-Brexit agreement dealing essentially with the immediate implications of 'taking back control' of the fishery resources within the UK's EEZ and the likely shape of the

DOI: 10.4324/9781003362913-11

UK fisheries policy in the immediate future. Then we finally venture further into the future unknown in considering how the separation might affect the overall management of what will still be the shared seas around the UK's coast and the UK's future relationship with the EU's CFP. But we begin our future analysis with a review of the circumstances that caused the seismic shift in political relations in the first place – that starts with a short recap on the 2016 referendum before proceeding to uncover the tangled web of the Brexit process up to the start of 2020.

Brexit: the political context

Brexit was a mistake – that is to say that the newly reinstated Prime Minister David Cameron, with a small but workable majority of seats, made a grievous miscalculation when deciding to hold a referendum on the UK's membership of the EU. Previous prime ministers had suffered embarrassment at the hands of a small but noisy coterie of Eurosceptic MPs. Cameron's position was no different except that, outside the main political parties, the anti-European interest in the UK Independence Party (UKIP) had found a strong, populist leader in Nigel Farage who was beginning to threaten the traditional allegiance of some of the more right-wing voters. Cameron felt the irritating itch and decided to scratch it through a showdown with the Eurosceptic wing of his own party in the form of a referendum to be held in June 2016. According to the increasingly unreliable opinion polls, he was in no real danger: the steadfast 'remainers' across the political spectrum would easily see off the challenge from the 'leavers'. Cameron was confident enough to allow his cabinet colleagues the freedom to support either cause.

What was to follow was the British electorate's first experience of a deluge of 'fake news, half-truths and empty promises' of life outside the EU and a pandering to the xenophobic mindset of some voters that was countered rather lamely by a flood of economic projections on the economic costs of leaving the EU that left most voters unimpressed and largely unmoved. The leadership on both left and right of the political spectrum had totally misread the public mood. There was little understanding of, and no great interest in, the substantive issues concerning the UK's membership of the EU. But there was a rising level of anger over the consequences of the previous ten years of apparent financial mismanagement and more especially the severe economic austerity that left some elements of the population and certain older industrial regions isolated from the benefits of a still prosperous Britain. All they wanted was the chance to bloody the noses of the political elites in London and Brussels whom they held ultimately responsible. The referendum provided just that opportunity and the 'sunny uplands' of an independent UK offered some imagined recompense.

The result, a narrow 52% to 48% win for the leave campaign, came as a shock to both sides. It was a verdict that gave the UK government no clear guidance as to what kind of Brexit – hard or soft – the electorate had voted for. Moreover, the narrowness of the verdict had left the UK deeply divided both regionally and politically. Two of the four nations that make up the UK had clearly voted in favour of remaining within the EU and crucially the Westminster parliament contained a majority of MPs who had voted to remain.

Worse was to follow. David Cameron immediately resigned as prime minister, and after a brief but tormented search, Theresa May was appointed as a 'safe pair of hands' to lead the Conservative government through the Brexit process. A remainer at the time of the referendum, the new prime minister made two serious miscalculations. The first was to call a surprise general election in 2017 in order to boost the slim Conservative majority, again relying on opinion polls that suggested that the Labour party, still recovering from the 2016 defeat and the controversial election of the renegade Jeremy Corbin as its new leader, would be in no condition to make a meaningful challenge. Although the Conservatives emerged as the largest party, the result was a 'hung parliament', reliant on the Northern Ireland Democratic Unionist Party's 10 MPs to secure the government's position, with an ebullient though still quite small Eurosceptic coterie – somewhat euphemistically calling themselves the European Research Group – left in a much stronger position to influence the government's European Strategy.

Instead of seeking the cooperation of the other main parties in developing a softer Brexit, possibly around the retention of membership of the single market or at least a customs union, that would avoid the worst economic consequences of Brexit and possibly even bridge the division between remainers and leavers, the chosen path was to appease the extreme Brexiteers through a hard line Brexit strategy that estranged the prime minister from certain implacable remainers within the Conservative parliamentary party. The final ingredient in the toxic 'witches brew' was the presence on the back benches of the parliamentary opposition of a significant number of remainers who had previously been returned to parliament by what had become leave-voting constituencies.

The substance of the initial Brexit negotiations with the EU was the drafting of the withdrawal agreement involving three key issues: the financial settlement, the status of the EU citizens choosing to remain in the UK (and vice versa) and resolution of the Northern Ireland question. Of these, the first two were dispatched quite quickly; the third, however, was to prove much less malleable. Indeed, squaring the circle created by the adoption of a hard Brexit, without retention of membership of the single market or a customs union, and the anomalous position of Northern Ireland as economically an integral part of the island of Ireland but politically an integral part of the UK, proved impossible. The sticking point was the need for a continued absence of any form of physical border separating Northern Ireland and the rest of Ireland as decreed by the Good Friday Agreement in 1988 that had ended decades of sectarianism violence within Northern Ireland costing many hundreds of lives. Even though it was widely anticipated in the accompanying political declaration that the issue could in part be resolved by the negotiation post-Brexit of a free trade agreement between the EU and UK, finding a solution to leaving without re-establishing a tangible if temporary customs border that satisfied a majority of MPs was to prove an unassailable problem.

Even though Prime Minister May had managed to negotiate both a withdrawal deal and an agreed political declaration over the UK's future relationship with the EU, the assent of Parliament was always going to be the problem and the voting behaviours of some individuals and informal groupings were likely to prove crucial to the outcome. The result was predictable: parliamentary stalemate and a failure three years on from the referendum to reach a vote enabling the UK to quit the EU. After two extended deadlines for leaving, the new incumbent Prime Minister, Boris Johnson, decided to call a snap general election. At the time, this appeared a risky proposition with the prospect of another hung parliament, months if not years of further negotiations and continued uncertainty for most sectors of the UK economy, including fishing.

In the event, it proved to be the right decision. Boris Johnson totally outmanoeuvred his opponents, judged the mood of the English electorate to perfection and turned it into a single issue – 'getting Brexit done' – election. The result brought closure to the question of whether Britain would be leaving the EU and under what conditions. With a very comfortable overall majority, the government had a clear mandate to fulfil the outcome of the 2016 referendum. It could move on quickly from completing the withdrawal agreement and begin the real task of 'taking back control'. The UK entered a relatively brief transition period when it would remain subject to the rules and enjoy the benefits of the single market, pending the finalising of a new relationship including the all-important free trade deal. There was, however, considerable doubt as to whether detailed negotiations on the new relationship could be completed by the scheduled date of 31st December 2020. Although there were provisions for a one- or two-year extension, a decision to add a further clause to the withdrawal agreement making it illegal to extend the transition period beyond 31st December 2020, resurrected the fears of a very hard, no deal Brexit.

Contrary to the views of some commentators, fisheries had not featured prominently in either the referendum campaign or the process of securing the withdrawal agreement. True, they were cited as an example of the benefits of leaving and photoshoots of a flotilla of fishing boats sailing up the Thames to the Houses of Parliament did add some colour to the media coverage at the time of the referendum. But the fishing industry's solid support for the leave campaign made little impact on the overall outcomes. Indeed Scotland, responsible for around two thirds of UK fish landings, voted strongly in favour of remaining within the EU. As a general rule, the debate over membership was not so much concerned with substantive issues relating to the UK economy as with the less tangible concepts of sovereignty, independence, 'taking back control' and whose responsibility it should be to make the laws that govern our economic and social life. Where discussion did turn to more substantial issues, it was more likely to be concerned with the UK's 'globalised' manufacturing sector and especially the car industry dependant on the import of key components from European partners and the need for frictionless trade to secure 'just in time' deliveries to UK based assembly plants. Nonetheless, the fisheries sector was expected to be significantly impacted by Brexit and one of the few areas of the UK economy to benefit directly from leaving the EU.

Fishing: the complicated issue of 'taking back control'

For fisheries, the real tasks of 'getting Brexit done' and 'taking back control' were about to begin. Potentially Brexit should be able to provide the UK with a rare opportunity to expand its fisheries sector quite substantially. But difficult times lay ahead. It would require considerable ingenuity and skill in negotiating the rather challenging route to achieving such a goal. Very little was known for certain as to what 'taking back control' actually meant for the future of UK fisheries. At this point, we should remind ourselves that in reality the phrase was for fisheries a delusion: the UK was entering a wholly new situation. There was no 'status quo ante' to which the UK was returning. Prior to the accession to the EU in 1972, the only exclusive fishing zone the UK could lay claim to was the 12 nm wide territorial seas around its coasts – of which the outer 6–12 nm zone was subject to agreements granting access to several third party states for specified species. Beyond the 12 nm limit lay international waters open to all comers.

'Getting Brexit done' as far as fisheries were concerned involved two quite distinct but interrelated tasks: the allocation of fishing rights within the UK EEZ and securing favourable access to European markets that were by far the most important current destinations for UK exports of fish and fish products, on which significant sectors of the UK fishing industry depended. In terms of the negotiating process, the UK and EU industries had very different views on how these two tasks should be tackled: the UK favouring entirely separate negotiations and the EU arguing that the two issues were interdependent and should be treated as one. The former would give the UK greater freedom in deciding how much of the EEZ's assets should be allocated to its former EU partners and under what conditions. The latter approach might help to moderate the UK's ambitions to expand its own fishing opportunities at the expense of those former partners for fear of losing unfettered access to its European markets. In the event, the fisheries negotiations were scheduled for completion by 1st July 2020, in part to allow for detailed discussions on the allocation of TACs for around a hundred or so 'shared stocks' in time for the start of the first post-Brexit fishing year in January 2021.

A further intriguing feature, and a possible complication, was that while the UK was in the driving seat over the allocation of fishing rights within its EEZ, it was the EU that would have the final say on access to EU markets as part of the wider trade relationships.

Dividing the spoils

Part V (Articles 55–75) of the United Nations Convention on the Law of the Sea (UNCLOS III, 1982) sets out the legal guidelines for the administration and management of the EEZ, making it abundantly clear that the coastal state assumes full responsibility in respect of 'exploring and exploiting, conserving and managing the natural resources whether living or non-living' (Article 56). It also requires the coastal state to 'determine the allowable catch of the living resources in its exclusive economic zone ... taking into account the best scientific advice available to it' and deploying measures 'designed to maintain and restore populations of harvested species at levels which can produce the maximum sustainable yield' (Article 61). Thus far UNCLOS III is unambiguous as to the coastal states' sovereign power and responsibilities under the 'new' geopolitical order.

However, it also confers certain obligations on the coastal state in respect of the historic claims of other neighbouring coastal states: Article 61, referring to the optimum utilisation of the EEZ's living resources stipulates that 'where the coastal state does not have the capacity to harvest the entire available catch, it shall ... give other states access to the surplus of the allowable catch' taking into account 'all relevant factors including the significance of the living resources of the area to the economy of the coastal state concerned ... and the need to minimise economic dislocation of states whose nationals have habitually fished in the zone'. Such access would be subject to compliance with the donor state's licensing, management and enforcement protocols. It is thus clear that under Brexit, the UK's previous EU partners would have very sound, historic bases for exercising their claims to shares of the surplus allowable catch.

Where UNCLOS III is less helpful is over guidance on how the coastal state should determine the surplus allowable catch – whether on the basis of existing catch records or on present and future plans to expand its own harvesting capacity – and also over the vexed questions of 'shared stocks' and rights of 'geographically disadvantaged states'. Article 63 acknowledges the awkward situation where 'the same stock or stocks of associated species occur within the exclusive economic zones of two or more coastal states'. Likewise, Article 70 recognises the rights of geographically disadvantaged states 'bordering enclosed or semi-enclosed seas whose geographical situation makes them dependant upon the exploitation of the living resources of the EEZs of other states ... for the nutritional purposes of their populations'. Neither Article offers solutions to these dilemmas.

At this point, it becomes necessary to forsake UNCLOS III, designed to accommodate the basic principles applied to the global situation, and turn instead to the particular circumstances concerning the political geography of Brexit, superimposed on the already irregular and fragmented geography of Europe's coastal periphery. A glance at the map describing the political geography of post-Brexit marine territories is sufficient to demonstrate the complexity of the challenges facing those responsible for solving the fisheries question. The most obvious feature is the glaring hole created by Brexit destroying the geographical integrity of the 'common pond' that had previously underpinned the logic of the CFP.

Next, the highly irregular nature of the semi-enclosed spaces that make up the 'common pond' suggests that once a new political boundary is drawn through the middle of these semi-enclosed spaces the problems of 'shared stocks' and 'geographically disadvantaged coastal states' are bound to multiply and intensify. The protection previously afforded by the CFP to both fish stock management and the fishing interests of small coastal states like Belgium and the Netherlands is seriously compromised. New bilateral agreements similar to that overseeing the determination of TACs for shared demersal species in the northern sector of the North Sea, involving the EU and Norway, may have to be replicated elsewhere in the North and Irish seas and in the English Channel to solve the recurring problems of shared stocks. Moreover, future collaboration between the UK and EU will be required over the granting of access to fishing opportunities for the disadvantaged Belgian and Dutch fishing industries. So much for 'taking back control' and dispensing with the CFP!

Finally, the very uneven distribution of the benefits from reclaiming the fishing rights within the UK's EEZ becomes self-evident. The UK as a whole is to some degree disadvantaged by the fact that only over a few limited stretches of the EEZ do the boundaries extend out to the full 200 nm (Figure 8.1). In the English Channel, the median line separating UK waters from those of its neighbours barely reaches the 20 nm, and, in some brief instances, the 10 nm mark. As a result, Scotland can lay claim to the lion's share of the EEZ, leaving England with a significantly smaller fishing zone and both Northern Ireland and Wales with quite limited areas of the EEZ. In certain less advantaged areas like the English Channel the fishing industries were looking for a solution that would increase rather than close down the prospects of claiming further access to fishing grounds outside the national EEZ.

Assuring access to markets

Working out sensible solutions to the issues of future access to resources within the EEZ was only half the battle of ensuring that the UK secured the best possible outcome to Brexit for all its fishing-related activities. A second front was opened up to deal with future trade relations between the UK and the EU in the hopes of acquiring a free trade agreement. Unlike access to resources, where fishing interests take centre stage, in this much larger and far-reaching campaign the fishing industry was little more than a passive bystander. Failure to deliver a free trade deal would, however, reduce the much-valued 'taking back control of our waters' to a rather hollow Pyrrhic victory for some sectors of the fishing industry if tariff and/or non-tariff restrictions were to impede UK access to the largest and closest markets for exports of fresh fish and shellfish produce.

Over the past 50 years the UK seafood supply chain has profoundly changed. From the time when in the 1970s Britain lost its access to the most distant

FIGURE 8.1 UK EEZ (with nominal boundaries for England, Wales, Scotland and Northern Ireland)

water fishing grounds, it has ceased to rely on landings by UK vessels for its viability. The void left by the demise of distant water fishing was quickly filled by imports of fresh and frozen cod and haddock mainly from Iceland and Norway. Meanwhile, the UK catching sector was gradually becoming more closely aligned to the opportunities affected by the large, expanding European common market for sales of high-value fish and shellfish. The result of these changes is neatly encapsulated in the oft quoted shorthand description of the fish trade as 'much of the fish that Britain eats is imported and most of what it catches is exported'.

Over a similar time frame, the nature of what was being traded has changed. Although a significant share of current exports involve bulk transfers of pelagic herring and mackerel in the demersal sector what had formerly been seen as a relatively low-priced bulk commodity was now being marketed as a more finely differentiated food product, valued for its nutritional content and with a greater emphasis on quality assurance and certified environmentally sustainable origin, that needs to be handled carefully to avoid spoilage. Such a produce would expect to command higher retail prices.

Just as fisheries policy is focused on sustainable harvesting of the resource with little or no concern for what happens to the fish once it leaves the quayside market, so too the fishers' interest usually ends when the fish is sold. Despite attempts to bring fishers, processors and retailers together in discussion groups so they might better understand each other's challenges and strategies within the seafood supply chain, the fisher often feels estranged from any sense of further involvement. In the early days following the Brexit referendum, certain fisher organisations tended to be dismissive of the relevance of future trade talks; all that really mattered was ensuring a significantly greater state of the catch in UK waters. As the situation became clearer this instinctive reaction was tempered by a reluctant acceptance that the outcomes of trade negotiations could have greater relevance for certain sectors of the industry. Priority still lay with the balancing of shares in the TACs for UK waters. Significantly, the industry was anxious to gain assurances from the government that access to the resource and access to markets would be dealt with separately in the upcoming negotiations. Such assurances were willingly given by Michael Gove, then Secretary of State for the Environment, and frequently reiterated, though they began to sound increasingly hollow amid growing speculation that fisheries might again become a pawn in the much bigger game of resolving future trade relations between Britain and the EU.

In reality, anything less than a comprehensive free trade deal would prove challenging for some parts of the UK's fisheries-related economy; and that outcome was made to appear less likely by suggestions in official quarters that the UK no longer wished to retain full regulatory alignment with the EU. The extent of the damage would depend on whether a compromise deal was agreed or a no-deal Brexit leading to the adoption of WTO rules was the outcome, but anything that interfered with the notion of a 'frictionless trade' would pose potential problems. Friction is caused by the introduction of tariffs on fish and fish products exported to the EU and/or requirements for substantial additional documentation relating to each consignment.

If negotiations failed to deliver a free trade agreement, the options would have been limited to trading as an EU 'most favoured nation' or under WTO rules. In either case, the impact would be quite severe for enterprises reliant on the export market. According to a report prepared for the UK Trade Policy Observatory at the University of Sussex, even under favoured nation status, tariffs for fish would be set quite high varying from 16% on shellfish such as Nephrops and scallops to 7.5% for haddock and saithe.

While it might seem logical for tariffs to pose the greater threat, in practice the higher risk could come from the imposition of non-tariff measures and the swirl of documentation that must accompany each consignment. Fish, particularly when traded in fresh form, will attract more detailed scrutiny at the port of entry, especially in cases where a single consignment may include a range of species with different certification requirements. The resulting risk of delays could prove crucial; delays of two or three hours would be enough to disrupt delivery schedules and compromise the quality of the fish, thus prejudicing future contracts.

Key issues for negotiation and decision

The most persistent thorn in the side of the UK fishing industry is the presence of a large number of foreign vessels in what the industry has long considered to be 'our waters' with a license to harvest a significant share of 'our fish'. With the decision to take back control of our waters, the industry's expectations of a much brighter future were running high, driven in part by the longstanding grievance of unfair treatment at the hands of the European Commission.

Further fuel was added to the flames of apparent injustice by Ian Napier's detailed analysis of EU fisheries statistics for 2012–2014 that graphically revealed the extent of incursions by non-UK member states into what would become the UK's EEZ and the remarkably high levels of dependency of their fishing industries (and their domestic seafood supply chains) on access to these waters. Overall, of the average annual 1.1 million tonnes of fish and shellfish caught within the UK's EEZ, 58% were taken by non-UK vessels. When landing value is considered the balance is reversed: UK vessels accounted for 57% of the average £950 million of earnings and other EU vessels 43%.

Closer scrutiny reveals significant variations in the types of species caught. Non-UK incursions were weakest in the case of shellfish (14% of catch by weight) and strongest in the case of industrial 'trashfish' (80%) and pelagic species (61%). Similarly, with regard to the geographical distribution of non-UK fishing interest, not surprisingly the North Sea – on the doorsteps of Denmark, Germany, the Netherlands and Belgium – was the favoured destination. For reasons of geographical proximity, both France and Ireland preferred to fish UK waters in Area VII, stretching along the length of England's south coast from the Straits of Dover westwards into the Celtic Sea, principally for high-value demersal species.

What was given less prominence in Napier's analysis was the high levels of dependence on opportunities in what was to become the UK's fishing zone affecting the fishing fleets and seafood supply chains of several EU member states. Unsurprisingly Belgium, with its short coastline and reliance on UK waters for around half its annual catch, was likely to be the most 'geographically disadvantaged' coastal state as a result of Brexit. Five others – Denmark, Germany, Netherlands, France and Ireland – all with roughly a third of their landings coming from UK waters would be significantly impacted. For all six cases, a major rebalancing of access rights could cause serious dislocations to their fishing industries. It was hardly surprising, therefore, to find their fishing organisations and politicians urging the European Commission to take a very firm stance in the negotiations over the fisheries question; and every reason for the UK government to give very careful consideration to these concerns before deciding its own position.

Without doubt, therefore, the key question for both parties in addressing the fisheries questions was just how much access should be granted to neighbouring EU states in respect of fishing opportunities within the UK's EEZ. Not only would that decision define the UK's future relations with its neighbours over fisheries matters for decades to come and determine how easy it will be to cooperate with the EU over shared problems of marine policy and management, but it would also provide the key to answering many other queries relating to the fisheries question.

In an initial response to Brexit, Professor Jeremy Phillipson and I counselled a cautious, conciliatory approach to this issue by juxtaposing 'good' and 'bad' neighbour scenarios. The latter would seek to 'nationalise' the resources within the EEZ, clamping down firmly on access by third parties to its fisheries – an approach likely to create a reciprocal hard-line response on future trading relations in respect of fish and fish products. It would also generate a much less congenial environment in which to resolve issues of mutual concern in respect of managing fisheries in areas like the North Sea. A worst-case scenario could involve a complete breakdown of viable management strategies for shared stocks and a return to serious levels of overfishing of several key species. By contrast, a 'good neighbour' approach would look to achieve a more modest rebalancing of fishing opportunities phased over a ten- or twelve-year period to allow both the UK and EU industries time to adapt to the changing circumstances before the process is completed. This, in turn, should raise expectations of a less hostile environment for trade negotiations and the long-term relationships for shared management responsibilities in the seas around Britain that would inevitably follow.

We identified several other issues that needed to be addressed. Taking back control raises immediate concerns over the UK's monitoring and enforcement capabilities. The sheer length of the boundary passing through five seas demarcating the EEZ will make it difficult to police the UK's fishing zone under normal conditions, even with a doubling of the number of Royal Navy patrol vessels, increased air reconnaissance and sophisticated satellite technology; and impossible in the unlikely event of a well-planned 'mass trespass'. Accurate recording of the catches taken by non-UK vessels within the UK fishing zone could also prove problematic.

Other questions related directly to the form in which the rebalancing would take place – either as a simple fixed share of the TACs across all species or in a more tailored form varying the level of shares according to specific species creating greater benefits for the UK fleet or, conversely, seeking to minimise the damage inflicted on the EU member states. The assumption, in either scenario, would be for the EU to be wholly responsible for decisions on how the quota cuts would be distributed among its member states while member states would determine which vessels would be licensed to fish the revised quota.

There are a number of important questions relating to domestic issues concerning the allocation of any additional TAC within the UK. Should a portion of the additional TAC be held in reserve as an added safeguard for assuring maximum sustainable yields and/or for use in cases of emergency? How much of the extra TAC should be top sliced to provide the basis of greater viability for small-scale fisheries that have been starved of adequate quota in recent years? And how much should be set aside to support new start-up enterprises before distributing what's left among existing enterprises? Although such issues will have been given careful attention within Defra and Marine Scotland, no firm proposals had been made public by the start of 2021. Indeed, the answer to all of the questions in part depended on the scale of rebalancing to be undertaken.

Implications for the future of UK fisheries

The catching sector

Earlier in the chapter, we noted the unequal distribution of the EEZ among the four 'nations' that make up the UK which meant that fishers in parts of England and in Wales would gain relatively little from Brexit by comparison with those in Scotland. In theory, this should be of little consequence to all but the small-scale enterprises, confined to their local grounds, as the EEZ is not yet parcelled out among the four waters and fishers may fish any part of the zone providing they hold the appropriate licence and quota entitlements. In practice, the situation is rather different. Since the demise of distant water trawling, the centre of gravity for UK fishing activity has shifted from the Humber ports to the north-east of Scotland and the majority of fishing effort is confined to 'home waters'. Pelagic fishing for herring and mackerel is the principal exception, together with the high-value Nephrops and scallops targeted by vessels using trawls and dredges respectively. Thus Scotland, already the main focus of fishing activity, was expected to gain significantly from taking back control of our own waters, while fishers in parts of England and in Wales might have much less to celebrate.

Turning to the likely consequences of failure to secure a free trade deal, the impact was again expected to be distributed unevenly between different regions and across fishing sectors. Of the sectors at risk, the shellfish industry was likely to be the most severely impacted with around 80% of the annual harvest, worth £334 million, destined for export and the EU by far the largest market, accounting for circa 80% of the trade. The shellfish sector incorporates two distinct subsectors: wild capture fisheries, notably for crustacea such crab, lobster and Nephrops, and mariculture production of molluscs, mainly mussels and oysters.

Together these two sub-sectors have provided the main source of growth in the UK fishing industry over the past 35 years or so, creating much-needed alternative fishing opportunities for small boat operators in particular. That growth has been largely concentrated in England and Wales where shellfish are now responsible for 69% of total annual catch by value.

The reasons behind this expansion are threefold: stronger recruitment to adult stocks due to lower predation from declining demersal species; a surge in market demand mainly from the near continent; and exemption from EU regulations and restrictive quota. Only one shellfish species – the hugely important Nephrops now the UK's second most valuable catch – is subject to TAC and quota regulation. Otherwise, shellfisheries are subject to fairly tight regulation: in England, locally devolved management organisations known as IFCAs (Inshore Fisheries and Conservation Authorities), use local byelaws to control fishing effort and, in the case of mariculture production, regulating orders covering a designated area of inshore waters dedicated to mussel and oyster cultivation.

Geographically, the areas most likely to be affected by disruption to frictionless trade with the EU are located along England's east coast from the Tees to the Thames. Here, crab and lobster dominate landings at the small ports interspersed with major mariculture enterprises around the Wash. The pattern is repeated in smaller concentrations of shellfish production along the south coast and again in south and north Wales where mussel farming assumes local importance.

While on the subject of shellfish mention must also be made of the situation in the economically more deprived west coast of Scotland where the explosive growth of fishing for Nephrops, deploying both static and mobile gear, has seen the revival of fishing over the past two decades. Take the Outer Hebrides for example, where according to an independent report on the future of fisheries management in Scotland published in 2010, the fishing industry's dependence on shellfish in general and Nephrops in particular is extreme, accounting for no less than 98% of the total catch by value. The islands, already disadvantaged by distance from urban markets and with little else by way of employment to supplement the crofting income other than tourism and weaving on Harris and Lewis, would be very hard hit by the impact of tariff and non-tariff measures.

A similar, though less severe scenario confronts a significant number of small and medium-sized enterprises on England's south coast that have sought to capitalise on their locational advantage in respect of London as a domestic market and the Channel ports affording quick access to customers in Paris, Brussels and Amsterdam through sales of fresh fish at premium prices. Merchants in Brixham – England's principal fishing port in terms of landing value where roughly 60% of landings are destined for EU markets – have developed distribution systems guaranteed to deliver fresh fish to customers on the near continent within 24 hours of landing. Such systems are placed in jeopardy by the risk of serious delays at the busy Channel ports arising from the introduction of non-tariff restrictions in the form of complex documentation requirements.

In general, it is the small-scale enterprise that has the most to fear. It is less well placed to cope with such situations given the often diverse make up of small unit landings, lack of time and expertise in dealing with complicated paperwork, low profit margins, meagre financial reserves and little collateral. It also had the most to lose should it be forced to sacrifice premium prices for top-quality fresh fish intended for EU customers and be exposed to competition with much larger enterprises on the domestic market. Adapting to change of a magnitude posed by the ending of frictionless trade with the EU would be difficult, not least because of the apparent inelasticity of the domestic market in terms of consumer expenditure on fish, already a relatively high-cost alternative food product, and the rigidity of consumer choice over preferred species. The oft stated 'opportunity to open up new overseas markets' simply does not work in the context of small-scale enterprises in small volume sales of high-quality fresh fish.

The processing sector

There is one further aspect of the seafood supply chain that merits inclusion in a review of potential impacts on the UK fisheries sector. Processing covers a wide range of activities involved in preparing the fish before it reaches the consumer. Primary processing involves the basic tasks of heading and gutting the fish usually done on board the fishing vessel, together with filleting by merchants, before presentation as fresh fish for the fishmonger's slab. Different forms of processing, including freezing, salting, curing and drying, are used to preserve the highly perishable commodity and give it a significantly longer shelf life. Finally, secondary processing often transforms the product into oven ready convenience food. All processing is essentially concerned with maintaining or enhancing quality and so adding value to the end product. What is important to remember in the context of the present discussion is that a very high share of the fish handled by the major freezing and secondary processing firms, concentrated in northeast Scotland and on Humberside, is actually imported mainly from Iceland and Norway. Those firms could thus face the prospect of both tariff and non-tariff measures applied to imported raw materials and exported fish products.

Another important facet of Brexit could prove of even greater consequence for the UK's processing sector. It concerns the decision of the UK government to replace the previous free movement of people within the EU by much stricter controls on migrant labour, whether from the EU or other parts of the world, through a points-based qualification system that will in practice ban unskilled workers from entry. Currently, the processing sector is heavily dependent on contract labour originating mainly from eastern Europe. A survey of 18 Scottish processing firms in 2015 accounting for over 35% of the sectors' workforce, found almost 60% of workers were from non-UK countries, notably the Baltic states of Poland, Lithuania and Latvia. The offshore fishing sector in Scotland also relies on overseas labour for around 25% of its crew members. Such figures raise doubts about the industry's ability to expand its harvesting capacity in the event of increased quota allocations. But it is the processing sector that faces the greatest challenge. Failing some concessions from the government over future migration rules it will be hard pressed to attract sufficient workers from the local and regional labour markets to maintain production at existing levels without significant increases in hourly pay. High turnover of labour would add further to production costs. For those firms with little to tie them to the UK, one option might be relocation overseas where none of the potential Brexit constraints would apply. The consequences for the fishing sector and the domestic seafood supply chain would be far reaching.

The North Sea: a theatre of broken dreams?

Returning to the analogy of the theatre first introduced at the very outset of this volume, and as a means of completing the review of issues centring on Brexit, it is likely that the denouement of the drama will be played out with most eyes firmly fixed on the North Sea as the crucial backcloth. This remarkable little sea, once the epitome of Nature's bounty with its prolific and diverse fisheries, has become a focus of critical attention during the troubled history of the last 40 years or so. Underpinning the abundance of the North Sea is a hydrographic regime in which the circulation of colder Atlantic waters and its intermixing with the warmer nutrient-rich waters of the North Sea lays the basis for remarkably productive fisheries that have endured a long history of exploitation by some of the world's most industrialised and urbanised coastal states.

Over the past 40 years the North Sea has been exposed as one of the world's most heavily fished areas now suffering from depleted stocks – some in a very precarious state – and greatly reduced fishing opportunities. But this is not the only problem to confront the fishing industry. In the recent past, the area has experienced severe pollution, largely as a result of several major rivers in the UK and on the continent discharging contaminated waters from heavily industrialised and modern chemically treated farming hinterlands into the sea. It is also one of the most congested seas in terms of competition for space for the extraction of aggregates and hydrocarbon energy resources, navigation and, most recently, renewable energy generation. And its waters, primary inshore, are among the most highly protected in terms of wildlife conservation thanks in part to the successful roll out of the EU's Natura 2000 programme. All of these constrain the fishers' freedom to fish, but none is more limiting than the ever-increasing burden of regulations intended to protect fish stocks. It is doubtful whether any other part of the common pond has suffered more from the failure of the CFP in the late 20th century to arrest the decline in fish stocks and restore them to stable sustainable levels.

Looking briefly to the not-so-distant future, the impacts of climate change in terms of sea level rise and the changing make up and productivity of fish populations on the increasingly warmer waters could well prove a greater existential threat to North Sea fisheries. The impacts of global warming on marine ecosystems and populations of cold water fish species are likely to become evident rather earlier than expected with cod, haddock and other mainstays of the commercial fishing industry struggling to prosper except in the most northerly seas or in deeper waters off the coastal shelf. Early signs of the northward drift of cod stocks in the temperate waters off the west European coasts have been evident in the North sea for the past few decades and the sudden relapse in the recovery of North Sea cod stocks in recent years may prove symptomatic of what is to come. What is virtually certain is that over the next 20 or 30 years the geographical distribution of key commercial species in areas like the North Sea will alter, and 'tropicalisation' of hitherto temperate waters with increasing incursions of warm water species like red mullet, will continue. And with it the transformation of the UK's fishing industries.

Some of the more fanciful 'solutions' to the risk of widespread, permanent, inundation of extensive low-lying coastal areas caused by sea level rise pose an even greater threat. Recent suggestions to enclose the North Sea through a series of embankments connecting Norway, the Shetland Islands and the Scottish mainland in the north and between Land's End and Cape Finisterre in France to the south at the cost of around £400 billion, would turn the Baltic and North Seas into a vast 'freshwater' lake and by definition eradicate all forms of marine fisheries!

Returning to the present, and recapping some of the earlier findings, the reason for choosing the North Sea as the backdrop for the final act is the simple fact that the full impact of Brexit will be felt by all of its coastal states in terms of both winners and losers. As a result of taking back control of its fisheries, the UK faces the anomaly of being responsible for the major share of the North Sea but less than half its annual fish harvest. Redressing this balance seems both logical and necessary. This will, however, have serious consequences for the fisheries economies of the four EU member states bordering the North Sea. The Netherlands and Belgium both have their EEZs wholly contained within a small area of the North Sea, while the EEZs of Denmark and Germany do allow limited access to the neighbouring Baltic Sea. According to Napier's analysis, Belgium relies on access rights within the UK EEZ for around half its catch and is clearly the most 'geographically disadvantaged' coastal state as a result of Brexit. The other three coastal states have been dependent on access rights to what is now the UK fishing zone for up to a third of their annual catch by value. The fishing industries and seafood supply chain of all four countries could be severely dislocated and employment prospects in the coastal economies impaired by a hard-line Brexit where there is little scope for relocation of fishing effort to other areas of the common pond.

Further reading

As is only to be expected there is to-date no comprehensive published account of the impacts of Brexit on the fisheries and their management. Source material is therefore more widely dispersed and fragmented.

Department for Environment, Food and Rural Affairs. (2018). *Sustainable Fisheries for Future Generations*, White Paper, Cmd 9660, London.

Gasiorek, M., & Walmsley, S. (2018). *Fishing in Deep Waters*. UK Trade Policy Observatory Briefing Paper 21. Brighton: University of Sussex.

House of Lords. (2017). *Brexit: Fisheries*. European Union Committee 8th Report Session 2016–17. London.

Marine Scotland. (2016). *Scottish Sea Fisheries Employment 2015*. Edinburgh: Scottish Government.

Napier, I. (2016). *Fish Landings from the United Kingdom's Exclusive Economic Zone: By Area, Nationality and Species*. Port Arthur: NAFC Marine Centre, university of the Highlands and Islands.

Napier, I. (2016). *Fish Landings from the UK Exclusive Economic Zone, and UK Landings from the EU EEZ*. Port Arthur: NAFC Marine Centre, University of the Highlands and Islands.

Napier, I. (2018). *The Potential Value to the UK Fishing Fleet of Larger Shares of the Landings from the UK EEZ*. Port Arthur: NAFC Marine Centre, University of the Highlands and Islands.

Phillipson, J., & Symes, D. (2018). 'A sea of troubles': Brexit and the fisheries question. *Marine Policy*, 90, 168–173.

Scottish Government. (2010). *The Future of Fisheries Management in Scotland*. Report of an Independent Panel. Edinburgh.

Symes, D., & Phillipson, J. (2019). 'A sea of troubles' (2): Brexit and the UK seafood supply chain. *Marine Policy*, 102, 5–9.

United Nations. (1982). *Convention on the Law of the Sea*. UN General Assembly.

9

2020

Getting Brexit done?

Introduction

This chapter focuses on the events of 2020 – a year that can be seen as instrumental in defining a new relationship between the UK and Europe covering a wide range of issues and possibly establishing a much-altered system of governance for fisheries across a wide area of Europe's continental shelf. Its aim is to trace the sequence of events leading to the signing of a new deal describing future relations between the UK and EU in relation to fisheries, examining some of the main influences that shaped the content of the deal and, most importantly, assessing the outcomes in terms of unveiling the promised bold new dawn for the UK's fisheries sector.

The withdrawal agreement outlining the terms of the UK's departure from the EU was finally endorsed by the British parliament on 10th January 2020, marking the *de facto* beginning of a 12-month transition period and paving the way for the final round of negotiations, defining the future relationship between the UK and the EU, beginning on 1st January 2021. Formal talks began on 2nd March with the fisheries agreement scheduled for completion by the end of June and the content of the overall relationship to be finalised in October allowing sufficient time for the EU member states and European Parliament to debate the proposals and sign off the new deal by the end of the year. As often happens, the ideal timetable was sabotaged by a combination of external events, obstreperous negotiators (on both sides) and the dragging of heels until the final days when the political leaders (Ursula von der Leyen, President of the European Commission, and Boris Johnson) take over and magically narrow the considerable gap between the negotiated positions sufficient for a deal to be struck at the eleventh hour.

In truth we know very little about the conduct of the negotiations. Unlike the proceedings leading to the completion of the withdrawal agreement conducted

DOI: 10.4324/9781003362913-12

in a blaze of publicity due mainly to exposure through the Westminster parliamentary process and a Brexit-obsessed national media, the final round of negotiations were held *in camera*. There were no extended briefings given at the end of each session and surprisingly few 'leaks' from those present at the talks. Only in the latter stages when the negotiations became more overtly political did we begin to get a clearer sense of what was actually involved.

There can be very little doubt that the Covid-19 pandemic exerted a significant influence over the negotiations both directly and indirectly. Not only would the pandemic make the conduct of face-to-face negotiations more difficult, but the secrecy surrounding the talks was enhanced by the fact that – starved of hard information – the attention of the media had switched almost completely from Brexit to much more immediate concerns over the rapid advance of Covid-19 in both the UK and Europe and the uncertain handling of the resulting crisis. This was particularly true of the earlier stages of negotiation when comment in the media was scarce and mainly confined to the middle pages of the newspapers. We shall return to the more direct impacts of Covid-19 on the fisheries question later. As a result of the secrecy of the negotiations, we must rely very largely on the outcomes to judge their efficacy.

Fisheries: the early months

Very early in 2020, it became apparent that two issues would come to dominate the overall Brexit proceedings: securing a free trade deal and fisheries, two rather strange bedfellows. Free trade was an obvious choice, considered by many to be the crux of the negotiations and likely to have a profound effect on the economy; by contrast, fisheries is an insignificant sector of the overall UK economy accounting for around 0.02% of GDP. Pre-selecting fisheries as a key issue thus requires some further explanation. Its relevance is essentially symbolic, closely linked to the concept of sovereignty. Taking back control of the fisheries in the UK's EEZ, managing the stocks and allocating the fishing rights was potentially a powerful signal of what Brexit actually stood for and one of the relatively few areas where Brexit could deliver an unambiguous win–win situation for the UK.

The interval between the completion of the withdrawal agreement and the starting date for the final round of negotiations provided an opportunity for the two sides to define their opening positions for the deliberations that lay ahead. This took the form of public 'conversations' between the fishing industries and their respective government institutions in which the industry representatives set out their demands and the governing bodies responded positively with pledges to uphold the interests of their fishing industries throughout the negotiations. Firm promises were made, some of which were at the time patently unsustainable. It came as no surprise to find the EU adopting a very firm stance over fisheries, arguing for the retention of the *status quo* in terms of access to the UK fishing zone. By contrast, the UK's position was based on replacing the existing regime with a completely new approach grounded in the concept of zonal attachment to determine access to fishing rights and annual renegotiations over their allocation following ICES advice relating to the annual state of the stocks and the recommended TACs. This was doubly disconcerting for the EU member states as it seemed to imply not only continuing uncertainty over future access but also that non-UK fishing nations might in future be expected to 'carry the can' for any reduction in TACs recommended by ICES.

The purpose of negotiation is, of course, to reach an agreed compromise from which the two parties can claim a partial victory and yet neither party is fully satisfied by the outcome. In this instance, the two parties were so diametrically opposed that it was difficult to see how they might be reconciled. There was also one further significant point of difference separating the two sides with the potential to disrupt the conduct of the negotiations. As elaborated in the previous chapter, the EU was insistent on the two key features of the fisheries question – access to fishing rights and access to the EU market for fish and fish products – being treated as contingent issues whereas the UK was equally determined to keep them completely separate. Over the course of the next ten contentious months, it was never quite clear as to which strategy prevailed.

Were we to seek guidance from some higher authority as to where the balance of probability lay in terms of a sensible, just and practical solution to the fisheries question, the UNCLOS III guidelines are perhaps the most appropriate. These clearly endorse the UK's claim of responsibility for future decision-making in relation to both the management of the living resources within the EEZ and for the allocation of fishing rights (quota) therein. By contrast, the EU's case rested heavily on the weight attached to long-established historic fishing rights, hitherto unchallenged, and the less clearly defined 'obligations', sited within UNCLOS III, and placed on the coastal state to 'minimise economic dislocation on States whose nationals have habitually fished in the [EE] zone'. No doubt the UK was emboldened by such an interpretation of international law when assuming a rather aggressive opening stance, but the strength of that position would only be truly tested when the time came to translate its obligations into actual figures setting out allocations.

Getting down to business (i): the early months

Once negotiations began, it soon became evident that four months was far too short a time in which to resolve the fisheries question. Progress in this period was glacial, hampered by the Coronavirus with discussions held at arms' length for much of the late spring and early summer months, and very probably frustrated by the enormity of the wall separating the two sides with no loose bricks visible that might enable one side or the other to gain some purchase on the discussions. There is no public record of the number, duration and content of the sessions dedicated to the fisheries question. In all probability negotiations began with the two parties talking in different tongues – the UK using the language of 'sovereignty' and zonal attachment and the EU negotiators replying in the language of 'relative stability'. As the completion date of the end of June approached, the impression was that little progress had been made in moving the discussion from first principles to substantive proposals. Both parties publicly blamed the other, with the UK complaining that the EU was in denial over the shift of responsibility and decision-making contingent on Brexit and obstructing progress by continuing to make 'impossible demands' based on their *status quo* stance. For their part, the EU pointed to the fact that further progress was being denied by the reluctance of the UK to table substantive proposals as to how the new decision-making might work or provide any details as to how fishing rights in the UK EEZ would be impacted.

It came as no surprise when the end of June was reached without resolution of the fisheries question and negotiations were extended for almost another four months before the new final date for submission of agreed proposals became due.

Lessons from the Covid-19 crisis

Events in the first half of 2020, linked to the Covid-19 pandemic, exposed some of the dangers facing a fishing industry heavily dependent on export markets that could both cripple the harvesting sector and undermine the domestic food chain when access to the overseas market is restricted. The lockdown of the hospitality sector both at home and abroad covering hotels, restaurants and pubs, was largely responsible for a sharp drop in demand for high value fish and shellfish, leading to the virtual closure of export markets. A broad swathe of the UK fleet reliant on export sales for their revenue were forced to rethink their fishing strategies. Significant numbers of small-scale enterprises simply gave up fishing with some putting their boats up for sale, while the larger scallop dredgers and Nephrop trawlers were obliged to tie up their vessels for the duration of the lockdown.

The domestic market was also affected. The lockdown of the hospitality sector and the decision by many supermarkets to close their fresh fish counters meant two of the most important channels for sales of fresh fish to the British consumer were out of action. Supply exceeded demand, prices slumped and those in the harvesting sector catering primarily to the domestic market were similarly obliged to adapt to the changing situation. For many faced with a fall of up to 50% in the price of some species, it was no longer worthwhile going to sea. Larger offshore vessels with high operating costs tended to replace the customary trips lasting several days with shorter trips in coastal waters. Most of the coastal fleet cut their number of days fishing. In Scotland, the absence of foreign buyers from quayside auctions and the lack of demand for frozen fish from regular institutional buyers further depressed the market. An agreement was reached in May between the POs and Marine Scotland for whitefish vessels to tie up for eight consecutive days in the hopes of rebalancing supply and demand and stimulating a recovery in prices.

Elsewhere there was an expansion in direct sales, bypassing the increasingly jittery quayside auctions, with local 'catch of the day' deliveries organised by individuals or local collectives of small-scale fishers. At least one larger, regional distribution system covering parts of southern England was initiated by a firm buying directly from a number of boats, grading, packaging and distributing the fish to merchants, fishmongers and individual customers with agreed prices set on a weekly basis for both the fishery enterprises and the customers. In Scotland, where domestic sales usually account for only a fifth of the Scottish catch, businesses were reported to be looking to increase domestic consumption and extend the search for markets beyond Europe. Both approaches face significant challenges, first in the re-education of the domestic palate and secondly in the future stability of overseas markets in what might turn out to be a deglobalising economy.

The impact on the UK harvesting sector was severe. Provisional data indicate the value of landings by UK vessels at UK ports in 2020 was down by 24% on the previous year – the result of a reduced volume of fish landed and a sharp fall in quayside prices (*Fishing News*, 11th March 2021). As expected, the shellfish sector was hardest hit with the Nephrops catch value down by 46% and that for crab by 42%. The value for demersal landings was also badly affected (−15%) with all main species showing significant declines in volume and quayside prices. Only the pelagic sector recorded an increase in the value of its landings, due mainly to a strong recovery of the mackerel fishery.

Developments arising from the exceptional events associated with the Covid-19 pandemic have a particular relevance for the future development of the UK fisheries sector. While the lockdown of the hospitality sector may be a temporary expedient, prospects for a return to normality with regard to the full recovery of the crucial export markets are less certain, due not only to the need to adapt to a post-Brexit future but also to a predicted global economic recession and the possible revival of campaigns for greater national self-sufficiency in food production and consumption that had begun to emerge during the early stages of the pandemic. Should such threats be realised, it could precipitate a more fundamental restructuring of the UK fishing fleet, processing sector and seafood supply chain.

Getting down to business (ii): the later months

It is unclear whether the tempo of negotiations increased significantly over the next four months and to what extent there was a sustained focus on the realities of building a stable, long-term future relationship over the region's fisheries. Some cracks were beginning to appear in the original facades to both sides' bargaining positions. Significantly, the EU was prepared to concede that responsibility for managing fish stocks in the UK EEZ would lie with the UK and that some reduction in EU fishing rights within the EEZ was inevitable, while the UK was willing to see a gradual programmed phasing in of the new regime rather than insisting on the annual renegotiation of fishing rights. But these concessions over first principles, though important, were insufficient to grease the wheels of serious negotiation. When values were eventually attached to the concessions, the two sides remained too far apart to be bridged by meaningful compromise. The EU was said to be willing to offer a reduction in fishing rights of between

12.5% and 15% over a period of seven years while the UK was rumoured to be demanding a 60% reduction over a much shorter three-year transition period. Such demands smacked of further posturing and a reluctance to engage in the business of resolving the fisheries question and creating a sound framework for fisheries management in the disputed zone. They also prompted questions as to the true goal of the negotiations: deal or no deal? Once again, the deadline for completing the fisheries negotiations passed with little evidence of substantive progress – and the same fate was to befall two other 'final' submission dates.

During the second half of 2020, the attitudes of the UK fishing industry seemed to alter, probably influenced by the lessons of the Covid-19-induced changes to the configuration of the seafood supply chain and as a reaction to the increasing probability of a no-deal Brexit as negotiations appeared to stall. Although still keen for the government to squeeze out the full benefits of taking back control of the EEZ, this was tempered by an increasing urgency to secure a trade deal that would ensure tariff-free access to established markets within the EU. Indeed, as the talks reached a crisis point in mid-December it was the Scottish industry's concern over the threat to exports, arising from a combination of tariff and non-tariff barriers, that dominated the headlines of *Fishing News*.

Early in November, the broader negotiations entered a new phase in which Boris Johnson and Ursula von der Leyen became more directly involved. Their first meeting in Brussels was perhaps more of a symbolic occasion ending in a pledge to redouble efforts to reach a deal over both fisheries and state aid rules, the two remaining stumbling blocks. A month or so later the two leaders renewed their discussions, this time ending by singling out fundamental differences over fishing as the principal cause for concern and with Boris Johnson subsequently putting the odds on a final settlement at no better than 50:50, adding that no-deal would still represent a 'terrific' starting point for a post-Brexit Britain. Whether this was the point at which the negotiations finally got serious or not, a new set of compromises tabled on 21st December created the glimmer of hope for a bridgeable solution with the UK seeking a reduction in EU catches within the UK's EEZ of a more realistic 35% – promptly rejected by the EU's chief negotiator, Michel Barnier – and a counter proposal of 25% from the EU. The following day von der Leyen took personal charge of the negotiations and on 23rd December a deal was to all intents and purposes concluded. Prime Minister Johnson had conceded to the EU's demands to limit transfer of fishing rights to 25%. Formal announcement of the result was delayed until Christmas Eve while detailed adjustments to the allocation of EU shares in individual fish stocks within the EEZ were completed.

Counting the cost of Brexit

For some time commentators had been warning the British public that a successful outcome to the final negotiations would yield only a thin deal at best. Having abandoned the prospects of remaining in both the Customs Union and the Single Market in 2017, a deal was always going to offer meagre recompense in terms of future ties with Europe. The decision to leave the Single Market – a concept nurtured by Margaret Thatcher – had meant sacrificing the considerable benefits of a paperless entry to EU markets post Brexit; leaving the Customs Union implied the loss of automatic rights to tariff-free trade with the EU. It was therefore incumbent on the UK negotiators to reestablish tariff-free trading conditions in order to guarantee preferential access to the UK's nearest and most important markets.

Resolving the fisheries question

The circumstances regarding fisheries were rather different. Despite concern over the future of the fisheries sector, the fisheries question had been completely ignored within the context of the withdrawal agreement, most probably because of its potentially toxic implications. It had, therefore, been left to the final round of negotiations to secure a long-term, sustainable resolution. The hopes of the UK fishing interests, bolstered by pledges to expand fishing opportunities as the catalyst for a revitalised fisheries sector, had been running high. However, fulfilment of such promises could only mean significant cutbacks in fishing rights for EU member states, sufficient to cause a major contraction of their fishing sector. The expectations of EU fishers were driven by fear of the consequences.

Ultimately, what confronted the fishing industries, already badly bruised by poor winter fisheries (2019–2020) and the disruption to seafood supply chains resulting from the Covid-19 pandemic, was a limited balancing of fishing opportunities and a somewhat desolate post-Brexit landscape strewn with the wreckage of broken promises and unfinished business. Representatives of the UK's catching sector were quick to accuse their government of a sell-out over fishing rights and a repetition of the betrayal in the late 1970s that had originally deprived the industry of its 'fair share' of fishing opportunities within its own EEZ (*Fishing News*, 14th January 2021). Seafood exporters, on the other hand, welcomed the free trade aspects of the deal but renewed their concerns over the altered logistics and overall costs of reaching their established EU markets.

The nub of the industry's complaints lay in the agreement whereby the *value* of the EU catches within the UK EEZ was to be reduced by 25% over a period of five years with an initial reduction of 15% in year one and by a further 2.5% in each of the four succeeding years. On the face of it, the deal had some of the hallmarks of the 'good neighbour' relationship described in Chapter 8 – a modest reduction phased in to cushion the impact – but the manner in which it had been ground out through protracted, sometimes confrontational negotiations had left both parties aggrieved.

However, the true scale of the changes only becomes apparent when the agreed reductions in value are translated into the actual volume of catches involved as demonstrated in the content of the Annexes to the main text of the agreement. As a result, the permitted share increases in catch levels for high-value demersal species are much smaller than for lower-value pelagic species. In real terms, for example, the gains at the end of the five-year period for those fishing the grounds off the south and south west coasts of England fall well short of those promised as a result of the application of 'zonal attachment' and will be very much smaller than the 25% national average while the share in the TAC for the UK EEZ held by UK fishing interests will be very little altered. Moreover, increases in UK quota shares for plaice and sole in the North Sea were based on so-called 'paper fish'. In each of the four years leading up to the 2020 negotiations, EU landings for both species had fallen well short of the ICES TACs and the agreed quota. Without a truly remarkable and immediate improvement in stocks, it was very unlikely that catches in 2021 and beyond would be any different. This would seem to suggest that the UK negotiators were either not fully conversant with circumstances surrounding the two species or, worse still, knowingly compliant in making what were, in the short term at least, empty promises.

Furthermore, the ending of the opportunity for FPOs to engage in quota swaps with their counterparts in Europe to remedy deficiencies in the formal quota allocation system and so allow a closer match between available fishing opportunities and the fishing profiles of both particular sectors and individual enterprises makes it all the more likely that they will be forced to suspend fishing activity before full uptakes of quota for key species, for want of sufficient quota for 'bycatch' species. An analysis undertaken by Marine Scotland, taking into account changes to TACs over the next five years, the profile of the Scottish demersal fishing industry and the impacts of ending both the 'informal' quota swaps and the benefits of the Hague preference scheme, suggests that the deal will negatively affect the sector. Hence the improbable headline in *Fishing News* (14th January 2021) that the 'Scots will have LESS fish to catch'.

Overall, the seemingly meagre benefits arising from the rebalancing of fishing opportunities were sufficient to raise doubts over the longer-term development of the fisheries sector and the levels of future private investment within the sector. Even the government's promised £100 m investment programme to restructure the industry and improve its infrastructural capacity was beginning to look precarious as its purpose was being redefined as compensation rather than developmental investment.

It was small comfort to UK fishing interests to learn that their European neighbours were also pouring scorn on the European Commission's handling of the negotiations that resulted in a sell-out of established EU fishing rights within the UK fishing zone. The resulting cuts to their entitlements within the UK zone would necessarily lead to a contraction in overall fishing activity and revenues with knock-on effects affecting their seafood supply chains and seafood security. According to the European Union Fishermen's Association (EUFA), the results spelled 'a dark day for the European industry', citing not only the immediate loss of fishing opportunities but also the failure to provide long-term security for the fishing fleets that would reopen the question of how to ensure sustainable management of the fish stocks. Closer to home, the leader of the

Killybegs Fishermen's Organisations in Ireland criticised the 'duplicitous nature of the protracted negotiations' leading to the shredding of repeated guarantees, singling out the implications for the important mackerel fishery that stood to lose out dramatically (*Fishing News*, 14th January 2021).

Analysing the text

For those hoping to see the negotiations lay the basis of a robust framework for a post-Brexit management system for the seas and living resources shared by the UK and EU, capable of bridging two quite distinct approaches and ensuring an equitable and sustainable exploitation of those resources, the outcome is profoundly disappointing. The surprisingly brief treatment of fisheries, covering a mere 14 pages of text and three annexes out of a document running to 1,240 pages, is confined to the very narrow tasks of agreeing the annual TACs and quotas within the UK EEZ. Chapter 3: *Arrangements on access to waters and resources* sets out the procedures for agreeing the TACs (Articles 6 and 7), with Articles 8–11 describing the steps required to reset the annual quota allocations for shared stocks as designated in Annexes 1, 2A and B.

In the final Chapter 4: *Arrangements for governance* provision is made for remedial measures (Article 14) including the option of suspending access to the fisheries and the introduction of tariffs on fish and fish products entering the EU markets. Of particular interest is the creation of the curiously worded Specialised Committee on Fisheries with the purpose of providing 'a forum for discussion and cooperation in relation to sustainable fisheries' and responsible for multi-annual strategies for conservation and management, and the management of non-quota stocks and the provision of emergency measures *inter alia*. Although the text offers no guidance as to whether it is to be an *ad hoc* or standing committee nor on its composition, the committee has the potential to exert a considerable collective influence over fisheries management in the post-Brexit era. Also included in Chapter 4 are provisions for the termination of the fisheries deal (Article 17) and for a joint review of the overall arrangements after four years and at similar intervals thereafter (Article 18).

Analysis of the text also reveals further evidence of the failure of the UK's ambition to take back a reasonable measure of control over its own waters, despite formal recognition of UK sovereignty and regulatory autonomy enabling it to develop its own management preferences on a non-discriminatory basis. Its inability to restore the integrity of the 0–12 nm zone – traditionally known as 'territorial waters' – in ICES areas 4c (southern North Sea) and 7d–g (off the south and west coasts of England), regarded as one of industry's earliest 'red lines', exposed the impotence of the UK's negotiating strategy. The result is of considerable concern to the small-scale sector already under-resourced in terms of access to quota-regulated stocks and still prevented from extending its low-impact, static gear fishing into the outer 6–12 nm zone because of the continuing presence of much larger, mobile gear, EU vessels.

The language of the text hints at a retreat from ground gained over the past 20 years in terms of more interactive, participative governance and a return to an essentially bureaucratic form of governance. There is an underlying sense of separation and suspicion between the two parties that necessitates the creation of mechanisms such as those described in Chapter 4 in order to secure agreement on the issue of future access to resources. Moreover, the opportunities for actor participation are significantly reduced following the loss of UK participation in the Regional Advisory Councils and the European network of FPOs and, at the scientific level, there are fewer chances to engage in collaborative research projects relating to the management of fisheries across Europe's continental shelf.

Finally, in analysing the text, there remains an air of uncertainty as to the duration of the agreement and the nature of the relationship post-2026, in part prompted by the juxtaposition of the two penultimate Articles 17 concerning termination and 18 dealing with review and renewal. Is the agreement to be seen as the core of a long-term solution to the fisheries question or interpreted as 'unfinished business' inviting a further round of negotiation to attempt a further extension of the UK's control over its own waters? The signs are that UK fishing interests are unwilling to settle for the 2020 agreement, while EU interests are determined to avoid further concessions. But would either party be prepared to entertain the prospect of another gruelling round of negotiations and risk the chagrin of facing the humiliation of another defeat, especially where the price of achieving only meagre gains was the imposition of tariffs and further disruption of the seafood supply chain?

An audit

Counting the costs refers not only to the outcomes of the final deal given that most of the disbenefits may be described either as 'invisible costs' – the emotional costs due to a sense of failure, loss of confidence and increased uncertainty – or latent costs such as those arising from the increased likelihood of 'paper fish' occurring particularly in mixed demersal fisheries where full take up of quota for key species may be denied by insufficient quota for other components of the catch. Other direct costs will be incurred as a result of the ending of the free movement of labour affecting the recruitment of crew and skilled workers in the processing plants. However, the more immediate concern comes from the non-tariff barriers relating to the highly complex new requirements involved in the export of goods to EU markets. For the fishing sector, the problem is exacerbated by the highly diversified and often small-volume consignments especially in the sale of wet fish whose on-time delivery is essential.

In the first month of the post-Brexit era, attention in the media quickly switched from dismay over the new deal to mounting concern over the true costs of trading with Europe and the chaos, delays and eventual rejection of some consignments destined for EU markets. Lack of preparation and experience in handling the additional documentation requirements at all stages, including the port authorities, lay at the root of the problem. Disruption to trade with the EU was deepened by the decision taken by some haulage firms to expand the so-called 'groupage' arrangements that allow several exporters to include their products in a single consignment. Notably, the fishing industry's biggest logistics provider DFDS called a temporary halt to groupage arrangements pending improvement to the IT interface and training of new staff. Meanwhile, the catching sector and coastal markets were held in a state of limbo. The hope was that such disruption would be short-lived, part of the inevitable 'teething process' as all links in the distribution chain learned from experience and confidence was restored. What is certain, however, is that trading with Europe will continue to face the issue of more 'red tape' and some of the costs involved will prove to be somewhat higher than expected.

There was one further twist to the events of 2020 that came as a reminder that there may be hidden bumps on the road ahead: the European Commission is now obliged to treat the UK as a foreign country and terminate the import of certain fish products on health grounds. At the end of January 2021, the Commission confirmed a ban on imports of live bivalve molluscs (including oysters, mussels, clams, cockles and scallops) not harvested in Class A waters or purified and tested within the UK so as to qualify for an export health certificate. Scheduled for implementation in April 2021, the new ruling was initially thought to apply only to wild-harvested molluscs and not the products of aquaculture. As most of the waters surrounding England and Wales do not qualify as Class A, most producers in England and Wales faced the loss of sales to EU customers worth millions of pounds with very little time to make the necessary provisions for purification.

For EU fishing interests the situation is very different. In addition to sharing most of the emotional costs and the additional costs involved in future exports to the UK, they face the palpable hardship of coping with the loss of 25% of the value of their catches in the UK fishing zone. For some member states, notably Belgium and the Netherlands, this will raise substantive issues concerning the re-structuring of their industry, the continued viability of individual enterprises and seafood security. Facing estimated losses of between €150 and 170 million each year, the European Commission's announcement of a €600 million compensa-tion fund will go some way to ease the immediate pain, but most probably prove inadequate in resolving the crisis facing fishing enterprises that had previously placed a high level of reliance on catches from the UK zone.

An inquest

It is too soon to pronounce a final verdict on Brexit. 2021 and beyond will be a period of trial and error for the fisheries sector both in adjusting to the conditions imposed by Brexit, in translating the instructions contained within the 2020 agreement into the foundations for a new stable relationship between the UK and the EU, and, over the longer term, in creating a shared vision for a sustainable and viable future for the fisheries and the means of securing it.

Certain provisional findings are, however, evident. The UK has failed in its ambition of taking back control of its own waters; there was no final resolution to the fisheries question; and there is as yet no clear mandate for future management of the seas around Britain. So why did it all appear to go pear shaped? The simplest answer is that such goals were never on the negotiators' agenda and the anticipated political context over sovereignty was essentially created for the benefit of the media and general public. Hopes on both sides were raised far too high in the run-up to the negotiations to be realistically achievable: disappointment was thus inevitable. Sceptics might also argue that fisheries has been pushed front of stage not because it embodied the notion of sovereignty and taking back control, not because it offered the opportunity for a win-win victory, but because it provided a significant bargaining chip in pursuit of a much bigger deal.

However, this alone cannot explain the failure of negotiations to realise at least some of the expectations of the UK's fishing industry. For this we must also turn to the unevenly matched negotiators and the tactics they employed. In Michel Barnier – a highly respected and experienced bureaucrat from within the Commission – the EU delegation had a truly professional and well-informed leader. By contrast, the UK delegation was led by a succession of politicians drawn mainly from the ranks of junior ministers in the Conservative government, lacking experience in the arts of international negotiation, poorly served in terms of detailed knowledge of the issues involved and seemingly ill-prepared for the task ahead. Their negotiating skills were severely limited and distorted by a blurred vision of post-Brexit, independent Britain, shrouded in confused rhetoric over 'sovereignty' and 'taking back control' that in a fisheries context was singularly difficult to substantiate.

In contrast to the EU's consistent objective of minimising any political damage from Brexit to its institutions, economy and the welfare of its 27 member states, together with its determination to stick to a clearly defined agenda, the UK's negotiating tactics appeared somewhat devious and desperate. Attempts to undermine the solidarity of the EU's position by seeking deals with certain individual member states that served only to strengthen the unity of the EU and later to provoke the EU into softening its stance with threats of withdrawal proved ineffective and self-damaging.

As depicted in the text of the agreement, the true brief for the fisheries negotiations was the simple task of reaching a workable arrangement over EU fishing rights within the UK's EEZ. Given this, the outcome was more or less predictable. The leverage afforded by the sheer weight of historic fishing rights within the zone and more particularly the Damoclean threat of 'no deal' enabled the EU to contain the UK's ambitions. Had the negotiations resulted in a larger transfer of fishing rights to the UK, might this have precipitated a stronger response from the EU fishing interests and an insistence on rejection? Probably not, as the EU also had more riding on a positive outcome to the overall negotiations than placating an angry fisheries sector. Had the inertia of the first four months been overcome sooner, would this have allowed space to be found for a more nuanced approach to the final agreement? Probably not. The die had been cast and the limited objective had been achieved.

Further reading

Heading Five: Fisheries, pp 261–274 and Annexes: Fisheries 1,2,3 in *Trade and Cooperation Agreement between the European Union and the European Atomic Energy Community, of the One Part, and the United Kingdom of Great Britain and Northern Ireland, of the Other Part* dated 24th December 2020.

Apart from the text of the Agreement, this chapter has relied heavily on reports in the media and on *Fishing News* in particular.

For an alternative, insider view of the negotiations, readers may find Barnier, M. (2021). *My Secret Brexit Diary*. Polity Press, both interesting and insightful.

10

BEYOND BREXIT

An uncertain future

Looking back

Our critique of Europe's fisheries policy during the turbulent years since c. 1970 – initially provoked by a forced retreat from distant water fishing grounds in the mid-1970s and sustained by the bold but flawed experiment of a common fisheries policy that became operational in the early 1980s – is nearing completion. There is but one task remaining: a tentative look into the future to see what lies ahead.

The CFP can hardly be hailed as an outstanding success except in terms of its durability. It presided over the continued overexploitation of EU waters and declining fish stocks during the first 30 years of its existence. Partially disabled at birth by its incompleteness and disfigured by the scars of compromise, its development was stymied by the promise of 'relative stability' embodied in the system of allocating fishing rights among the member states. Originally intended to ease fears over the surrender of sovereignty, 'relative stability' quickly became adopted by member states as the litmus test for future policy decisions. This narrowed the opportunity for fundamental reform, restricted progress to one based on incremental path-dependent change, and denied the CFP the flexibility needed to adapt to changing circumstances. Strangely, the European Commission has never quite managed to summon up the courage to challenge this profoundly debilitating condition.

Few member states would claim to be fully satisfied with the EU's handling of the fisheries. But none had sought to mount a sustained challenge to the CFP's basic principles and practices, not even the UK whose fishing industry had nursed a genuine grievance throughout the lifetime of the Policy. Blessed, in theory, with a relatively extensive, diverse and resource-rich EEZ, the UK found itself having to share this asset with other neighbouring member states in

DOI: 10.4324/9781003362913-13

a system that left the UK with less than half the available catch – a situation that became less tolerable as the resource base declined.

Recompense was meant to come as a consequence of the unexpected victory for the 'leave' campaign in the 2016 referendum on future membership of the EU that paved the way for the assumption of full control of the EEZ. The process of reaching that goal began with a long drawn-out struggle in the UK parliament to finalise the broad terms of the withdrawal agreement. Those terms were only settled after Boris Johnson's election as prime minister and a sweeping victory for the Conservatives in the general election of December 2019. And while this sealed the decision to leave the EU on fairly hard terms, working out the details of the UK's future relations with the EU was to prove a much more difficult – and eventually unsuccessful – challenge.

For the UK, 2020 marked the end of nearly 50 years of membership in the EU and almost 40 years of commitment to the CFP. That, in itself, marks a fundamental political transformation. Opting for a hard Brexit and, in particular, not electing to retain access to the 'single market' meant a virtually complete break with a system that had, in varying degrees, shaped the UK's economic, social and political development for more than a generation. As a result, the UK now finds itself the only nation in western Europe (and large swathes of eastern and southern Europe) beyond the direct reach of EU governing institutions.

Looking ahead

The implications of this new independence are difficult to grasp. It is the task of this penultimate chapter to sketch out some of the key factors likely to reshape the future governance and management of the fisheries sector, mainly in relation to the UK. That future is, by definition, uncertain and the further into the future we go the less information we have to guide our thoughts. For UK fisheries the uncertainties are magnified by the sense that 2020 had failed to establish a secure political future for the fishing industry and that the deal struck was most probably no more than a temporary stop gap.

2021: an unpromising start

Early in 2021 it soon became clear that, as far as the fisheries question was concerned, Brexit was far from done. True, the salient political issue of reallocating shares in the TACs for regulated stocks in UK waters between the UK and EU had been resolved, albeit to the satisfaction of neither party. But much of the heavy lifting associated with the detailed, essentially technical decision-making relating to the future management of the fishery resources within the UK EEZ was left to further negotiations between the UK and EU. There were three major concerns: management of the numerous 'shared stocks' within the EEZ; the resources within the UK's 12 nm zone, increasingly exposed to pressure from the introduction by certain EU member states of new, more efficient fishing gears;

and the implementation of the new arrangements for quota swapping between the UK and EU to ensure the best possible fit between established fishing patterns and available quota. In addition, but not involving the EU, there were crucial negotiations with third-party coastal states – notably Norway and the Faroes – concerning future access agreements in their waters for UK vessels.

Progress on all four issues during the first few months of 2021 was disturbingly slow. In fact, on matters relating to both shared stocks and inshore fisheries an early decision was taken to settle for fairly basic, provisional arrangements for 2021, deferring certain important key issues until 2022. More progress was made in respect of quota swapping where the go-ahead was given but the question remained as to whether the details of the new quota exchange mechanism to be finalised by the already busy Specialised Committee on Fisheries would be finalised in time to 'rescue' the 2021 fishing season for the all-important mixed demersal fisheries.

The reasons for the unpromising start to the new post-Brexit situation are complex but two circumstances may go some way to providing an adequate explanation. The first is the greatly exaggerated set of expectations as to the meaning of becoming an independent coastal state, resulting in an almost total lack of preparedness for the actual tasks that lay ahead. The second concerns the lingering sense that negotiations were still being strongly influenced by the contrasting political ambitions of the two parties. For the UK this meant striving to maximise so far elusive, tangible benefits of independence while for the EU it was an equal determination to minimise the further risks to the bruised status and integrity of the CFP. With neither party able to claim victory in the 2020 Brexit negotiations without clear evidence of substantive benefits (or meaningful damage limitation) for their respective fishing industries, there was little or no willingness to grant concessions in order to ease the path of the current technical negotiations.

The third-party negotiations over future access to fishing opportunities in both the UK and Faroese/Norwegian waters met more disastrous fates. Early decisions were taken to annul existing arrangements and make no provision for access in 2021, allegedly on account of excessive demands by the UK. This was a bitter blow for Britain's demersal fleet with an estimated loss of around 6,000 tonnes of cod in Norwegian waters alone. Once again, the UK's status and influence as an independent coastal state was being challenged and proved less effective than the long-standing respect for the EU's more moderate negotiating position that had earned the latter very favourable terms of access.

2021 is likely to have proved a difficult year for UK fisheries overall with the fateful combination of reduced fishing opportunities, weak landing prices and continuing problems relating to access to overseas markets. By no means all of the problems can be laid directly at the door of Brexit. As demonstrated in Chapter 8, the situation was complicated and made much worse by the impact of the Covid-19 pandemic on the food servicing industries throughout the UK, Europe and the wider developed world. Pelagic fisheries excepted, the combined effects of Brexit and Covid-19 have been felt throughout the UK's fisheries sector with the inshore sector in England and Wales most directly and acutely affected mainly as a result of the deep recession within food servicing across Britain and the EU and its slow recovery, but also reflecting unexpectedly high transaction costs incurred as a result of Brexit.

A good indication of the breadth of the recession in the UK's trade with the EU relating to the food sector in general and seafood in particular is provided by the Scottish Government's brief report, published in 2021, on *The Brexit Referendum 5 years on.* For the UK's food-related exports overall it was estimated that their value had fallen by £1.2 billion in the first four months of 2021 compared to the equivalent period in 2018 – a reduction of 34%. While the Covid-19 pandemic had clearly played a significant role, the direct effects of Brexit in terms of increased administration and transport costs, together with delays in the transit of goods to the EU were the focus of concerns within the food industry itself.

Circumstances relating to the seafood sector were broadly similar though the decline in exports over the same period was somewhat less severe at 27% but nonetheless indicative of a gathering crisis for the UK seafood supply chain and the harvesting sector in particular. The economic and social implications of rising transaction costs and declining market opportunities are considerable: an increasing number of firms are at risk of becoming economically unviable. Characterised by very large numbers of relatively small, independent family enterprises whose profit margins are inclined to be low and financial reserves limited, these risks are high. Occasionally, larger well-established firms that play a more central role in the local economy are forced to call it a day, as with a 40-year-old family business handling the export of landings from a significant number of lobster boats in Bridlington, one of the UK's leading shellfish ports (*Guardian*, 9th February 2021). But the vast majority of enterprises will survive the immediate sense of crisis more or less unscathed but with reservations as to what might lie ahead.

Where Brexit hurts most: the case of Ireland

Frustration, despair and anger over the implementation of the new fisheries agreement are not the sole prerogative of the UK industry. Breton fishers, for example, were incensed by the decision to exclude a third of their small-scale enterprises applying to fish in Jersey's territorial waters, mainly on grounds of insufficient evidence of having previously fished those waters. Similarly, the strict application of the new rules by UK authorities to limit the number of French vessels seeking to fish within the 6–12 nm zone off England's south coast has met with angry responses and threats of reprisal from both industry and government in France. Such actions are likely to make future negotiations over access and the management of both shared and unregulated stocks all the more difficult.

But it is Ireland, the UK's closest neighbour, that is hardest hit by the loss of fishing opportunities in UK waters. This is reflected in the allocation of the

EU's Brexit Adjustment Reserve (BAR), established to compensate the fisheries sectors and coastal communities in the six EU states directly impacted by Brexit and payable in instalments up to 2025 when the transition will be completed. Ireland is set to receive €1 billion or 27% of the total funding available, with the Netherlands placed second with 22% and France a relatively distant third (18%).

The final report of a seafood task force, established by the Irish government and charged with estimating the financial impacts of the 2020 fisheries agreement and recommending appropriate courses of action to mitigate those losses, identified three sectors of the Irish fishing industry as sustaining the most damage: the pelagic fisheries; Nephrops; and the mixed white fish sector. Of these, the fleet of 23 large, recently built, highly specialised freezer or fresher trawlers are predicted to suffer losses of €15.3 m in 2021 rising to €25.4 m by 2026 as a result of lost fishing opportunities in UK waters; a much smaller fleet of 15 less specialised pelagic vessels will incur losses of €1.5 m in 2021 rising to €3.1 m in 2026. Meanwhile, a fleet of 76 offshore trawlers is calculated to lose €4.2 m rising to €6.8 m by 2026. Finally, 52 white fish trawlers operating in ICES areas six and seven are expected to lose €3.6 m rising to €5.8 m by the end of the transaction period, though this could increase by a further €7.7 m should the resolution of the disputed Rockall fishery result in the exclusion of Irish fishing interests.

In terms of proposals to mitigate the impact of Brexit, the task force opted for three immediate courses of action, costing a total of €423 million to be funded through the EU's Brexit Adjustment Reserve. The first of these is a longer-term fleet restructuring measure through a decommissioning scheme aimed at removing some 60 offshore vessels from the active fleet at a cost of €75.7 m in order to restore the balance between fishing capacity and available fishing opportunities. The second involves short-term measures including voluntary tie-ups for a period of a month commencing in the final quarter of 2021 and payments to support the liquidity of four major fishermen's cooperatives and the processing sector *inter alia*, costing a further €70 m. But by far the largest course of action (€277.5 m) is directed towards 'onshore initiatives' to strengthen and enhance coastal communities dependant on the seafood industry. Included among these are capital investment in the processing sector to improve quality, efficiency and diversity; creating best practice, increasing efficiency and reducing environmental impacts in the aquaculture sector; the maintenance and modernisation of publicly funded infrastructure; and the expansion of local community-led initiatives designed to create alternative sources of employment, broadening and integrating economic activity and training and reskilling facilities to ensure the sustainability of coastal communities.

It is clear, therefore, that Ireland – along with the five member states bordering the North Sea – face varying degrees of reconstruction of their fisheries sectors and the provision of alternative job opportunities within their coastal communities in the immediate future if they are to remain viable. However, the extent to which the EU will need to revise the CFP, beyond the inclusion of the provisions of the latest agreement on access to fishing opportunities in UK waters, will not be fully known until the completion of the fourth decennial review on 31st December 2022.

Learning to live with Brexit (2021–2026)

The first few years of the post-Brexit era are likely to prove both difficult and crucial for the future development of fisheries in the waters of the North and neighbouring seas that host some of the most valuable fisheries in Europe. They will be difficult – especially for the EU – as the phased implementation of the agreement on revised access to UK waters is rolled out, culminating in a 25% reduction in fishing opportunities for EU member states and revealing the surprisingly meagre benefits that accrue to the UK fishing industry. By way of explanation for this apparent anomaly, it is important to remember that the UK industry was promised significant increases in fishing opportunities rather than in the overall value of the total catch. But, in fact, the 2020 agreement was framed in terms of value not volume.

These early years are crucial because unless there is a genuine closure by the end of the period, the future development of fisheries within these waters will in all probability be inhibited by further uncertainty. Should either party – most likely the UK with the 'unresolved' issue of the integrity of the 0–12 nm zone – decide to opt for further political negotiations to improve the situation, settlement of the fisheries question would be delayed several years.

For the transition to be completed with a minimum of stress and frustration, a change in the negotiating style will be needed from the basically defensive and at times combative approach to a more conciliatory one that focuses not on bruised political ambitions but on the welfare of the fisheries and those who live by fishing. Achieving and sustaining such a transformation could prove a delicate task, especially in a potentially febrile and less than fully stable environment: aggravation elsewhere in 'unfinished' areas of the Brexit settlement could seriously hinder a rapprochement between the EU and UK over fisheries. Ultimately, however, the longer-term futures for the fishing industries are likely to depend on whether the two parties opt for closure at the end of the transition and embed the 2020 agreement in their future fishery policies or choose to reopen negotiations in the remote hopes of achieving a substantially improved outcome.

Home affairs: policy change

Moving on to the less contentious issue of the UK domestic fishing policy, it was widely assumed that one of the key benefits of Brexit was the opportunity to develop a new, tailor-made policy designed to fit the needs of UK fisheries. This would replace the 'one size fits all' approach of the CFP addressing the requirements of no fewer than 15 EU coastal states within a fishing zone that stretched almost 5,000 km from the Gulf of Bothnia in the Baltic to waters off north west

Africa. A new national policy could make it much easier and quicker to change course should the relatively unstable conditions surrounding fisheries demand it.

A consultation document – *Sustainable Fisheries for Future Generations* – had been published in the summer of 2018 outlining the UK government's position. In it the government clearly declined the opportunity for a radical rethink of fisheries management, choosing instead a more cautious approach, pledging to continue its pursuit of devolved administration, an ecosystem-based approach and the goal of sustainable fisheries. This was scarcely an endorsement of the promised bright new future but rather the comfort of a 'business as usual' approach.

The new *Fisheries Act 2020* – the first major piece of fisheries legislation for over 50 years – passed into law in November, barely a month before the Brexit deal was concluded. It was generally well received throughout the industry with the CEO of the National Federation of Fishermen's Organisations (NFFO) hailing it as providing 'a balanced framework for a more agile, flexible system of fisheries management' (*Fishing News*, 3rd December 2020). It was however roundly criticised by marine environmental interests for its failure to include a number of House of Lords amendments concerning incentives for low-impact fishing and the absence of a binding legal duty to ensure that fishing for all stocks in UK waters was maintained at levels below MSY.

The Act was totally free of major innovations; indeed, in its content, it closely resembled the main features of the despised CFP. There were good reasons for this. First, as an independent sovereign state, the UK will face exactly the same challenges as those facing the CFP, with many stocks remaining on a knife's edge in terms of sustainability and likely to remain so for years to come. Secondly, there is no thoroughly tested alternative to managing the stocks than the present system of output controls using licensing and quota allocation, despite its reputation for being a rather blunt instrument. It was also clear from the direction taken in negotiating the new fisheries deal (Chapter 9) that it will be nigh on impossible to ignore the continued presence of the CFP in the management of the semi-enclosed, shared seas, requiring a high degree of congruence between EU and UK management systems. Finally, as it turned out, the continued complementarity of the two management systems will assist the process of implementing new arrangements concerning EU access to UK waters whose contrasting systems could further complicate the situation and provoke further dispute.

There is, however, one area of management that calls for urgent attention, namely the problem posed by the 'unregulated fisheries' occurring within the UK inshore zone. The anger and frustration at the failure to secure the exclusion of foreign fishing vessels from the 6 to 12 nm zone has been compounded as a result of the introduction, by certain EU coastal states, of larger and more efficient gears ostensibly targeting the unregulated species. While the exploitation of species like cod, plaice and sole are protected by TACs and quota, a significant and growing number of marketable species are unprotected except by local bylaws applying only to the inner 6 nm zone. Such species include a range of high-value shellfish – lobster, crab, scallop *inter alia*, mainly but not exclusively caught within the 6 nm zone, that feature among the top earning species in the UK, and the increasingly popular sea bass whose rates of exploitation are giving cause for concern.

There has been talk within both the EU and UK of action to control the exploitation of some unregulated fisheries but thus far no concrete proposals. The time is now ripe to translate talk into action both to ease the growing tension within the inshore sector and to secure the sustainability of the stocks concerned. However, the range of species likely to be affected will be small, simply because for most unregulated species, there is little or no scientific evidence on which to base the TACs and insufficient reliable data for the allocation of quota. Other means of control may be required, possibly including a more rigorous assessment and licensing of new types of fishing gear, such as fly dragging recently introduced by the Dutch, and possibly a modified form of effort control using the 'days at sea' mechanism. The task is by no means easy, and it will call for greater sensitivity and negotiating skills on the part of the UK and EU not only in formulating the regulation, but also in selling it to their respective industries, than has been in evidence so far.

Restoring confidence: the role of policy reform (2026–)

Disillusion presently runs deep pervading most areas of the fisheries sector. It involves a mixture of disbelief, anger and betrayal that promises of a brighter, more secure future for the industry has failed to materialise, together with uncertainty as to what the future may still have in store. But the despair is underlain by a deeper fear by many in the sector that those who manage the fisheries have little understanding of how the industry functions. Building confidence requires not only the essential tasks of tidying up after Brexit, including the provision of adequate funding to assist fishing enterprises to adapt to changing conditions, supporting the quest for new outlets for British fish and fish products both at home and abroad and agreeing protocols for the deployment of limited resources for policing the new agreement, but also a rigorous overhaul of the present approaches to the science and management of the fisheries.

An outdated science

In the context of our expanding knowledge and understanding of marine ecosystems and the implications for sustainable fisheries, it is increasingly clear that the current science of stock assessment and thereby our approach to fisheries management are archaic and therefore no longer fit for purpose. Several factors combine to explain and validate this provocative assertion. First, the basic approach to the tasks of stock assessment and deciding on TACs relies heavily on the analysis of recent trends in a procedure known as virtual population analysis (VPA) developed in the 1960s to predict the stocks' short-term futures. Despite the introduction of multispecies VPA modelling the structures of several interactive fish stocks, estimates are subject to fairly wide margins of error and with the adoption of the precautionary approach, the resulting TACs are likely to err on the side of under rather than overestimation of actual stock size. This helps to explain the industry's frequent complaint that the agreed TAC (even after its inflation during the policy process) falls short of what is commonly observed on the fishing grounds.

Unsafe TACs

With rather more damaging consequences for key sectors of the UK industry, ICES assessments and the ensuing TACs are essentially made on a species-by-species basis for each of the ICES areas, thus ignoring the basic principles of the now widely accepted ecosystem-based approach which argues that the sustainability of fisheries is dependent on the functioning of the ecosystem as a whole. It also takes little or no account of changes occurring in the distribution of certain species, including the northward drift of the cod. While the existing approach may yield relatively sound predictions for stocks that are targeted individually, it is becoming increasingly unmanageable for the mixed demersal fisheries that are a key feature of the Scottish industry.

Over recent decades, the innate difficulty of managing mixed fisheries has been made much worse by the instability of what had been the key species, cod. ICES proposals for 2022 TACs in the North Sea vividly illustrate the problem posed by marked imbalances between key components of the mixed demersal fishery, with cuts of 10% for the already much-reduced TAC for cod contrasting with huge increases of 154% for haddock and 236% for whiting (Fishing News, 8 July 2021). Changes of this magnitude render the management of the mixed fishery, where each individual haul will yield a variable mix of species, impossible. Early closure of the North Sea mixed fishery in 2022 already seems inevitable.

Although a dedicated committee on the ecosystem-based approach to management has been in place within ICES for a number of years, its findings have yet to exert any substantive influence over the conduct of stock assessments and the determination of annual TACs.

Lack of flexibility

A third, possibly less fundamental but nonetheless highly relevant, criticism of the current approach, adopted by scientists and tolerated by administrators, is its inevitable short-term future perspective and its lack of flexibility in the face of rapidly changing circumstances – most obviously in the context of climate change. At present the fisher's mind is focused on the present – today's and to-morrow's catch and the likely landing price for their next catch. There is little incentive to think ahead and consider changing their pattern of fishing behaviour because the opportunity for change is stifled by the largely intractable conditions governing their activity, with the licensing and quota systems the main culprits. It requires a crisis – like Brexit or Covid-19 – to think ahead and possibly consider a change of direction. There is growing evidence of significantly changing conditions in a broader set of environmental controls that govern their performance at sea – the dynamics and distribution of certain stocks, rising sea levels, warmer waters and increasingly unstable weather conditions that will, for the most part, impact unfavourably on existing patterns of fishing. The fishing industry will need more opportunities to adapt.

Taken together, these three issues present a formidable challenge to scientists, policy makers and those within the fishing industry itself. Everything seems to hinge on the willingness of fisheries scientists and ICES in particular to progress from reliance on the familiar single species assessment to a situation that combines these with a broader ecosystem approach in which the realities of interaction and interdependence of species and the significance of climate change can be more fully accommodated. Unless and until such a transformation is complete the goals of resource sustainability and optimising the economic and social values of fishing will continue to be frustrated.

Relations between the fishing industry and environmental interests

Not all of the problems currently facing the fishing industry can be laid at the doors of scientists and policy makers. The deterioration in relations between the environment lobby and the fishing industry over the last two decades is a case in point. It was triggered in the UK by the publication in 2004 of the Royal Commission on Pollution's controversial report entitled *Assessing the Impact of Fisheries on the Marine Environment* that recommended *inter alia*, reversing the then current presumption in favour of fishing, establishing a large-scale ecologically coherent network of marine protected areas covering 30% of the UK's EEZ as 'no take reserves closed to commercial fishing' and much tighter controls over fishing outside these reserves than those sanctioned by the EU. Although the UK government was reluctant to embrace such proposals within the timetables and standards proposed, the UK has witnessed a steady expansion of MPAs within its waters over recent years. And in 2020 the UK government recognised the concept of much more strictly protected zones, tantamount to 'no take zones', with the designation of 'highly protected marine areas' (HPMA) scheduled for 2022. However, the widening gap between environmentalists and the fishing industry owes more to the emergence, growth and increasing influence of more radical forms of environmental lobbying, aided and abetted by a supporting media and more in line with active public opinion, but well-grounded in the principles of the ecosystem approach. The contrasts in the language used and the actions proposed when faced with instances of overfishing and/or ecosystem damage are so great as to deny any likelihood of finding common ground. Radical environmentalism tends to demand the ultimate sanction involving the total exclusion of commercial fishing; the fishing industry is more inclined to find moderate solutions through selective action in curbing fishing effort through banning certain fishing gears, the introduction of seasonal closures and, less likely, a reduction in the number of vessels involved.

In one particular respect, the future looks quite bleak: the space available for fishing activity will be diminished. Further development of the MPA network in terms of area and content, together with the planned expansion of renewable energy development off the coasts of eastern England and Scotland will make significant inroads into the relatively rich coastal waters affecting both the inshore and offshore sectors of the industry, with little if any scope for accommodating the displaced fishing effort elsewhere within UK waters. Such developments make a detailed reassessment of the structure and nature of the post-Brexit fishing industry even more urgent.

Self help

Responsibility for modernising UK fisheries policy clearly lies principally with the relevant departments of the four constituent nations working together rather than separately. This does not, of course, preclude the industry from acting on its own initiative to bring about much-needed change. Throughout the past 50 years, the catching sector has, to its cost, remained isolated from and largely disinterested in the wider regional and local economies. It urgently needs to reexamine its roles at regional and local levels with a view to strengthening its position through greater engagement and leading to further diversification of its own activities.

Within the EU, the Fisheries Local Action Group (or FLAG) initiative, launched in 2008, has demonstrated how close cooperation between the fishing industry and the local economy can help diversify fishing-related activity, create new job opportunities and add value to the catch on a scale hitherto unexpected. The UK's initial response was reluctant, uptake was delayed and the eventual choice of locations was not always well researched, with the result that early returns were less impressive than in other EU coastal states – though enthusiasm for FLAGs was especially high among participants in the projects.

The emphasis of the EU FLAG initiative *per se* was focused largely on small-scale fisheries and supported by modest funding from the EFF and then EMFF, but there is no reason to suppose that broadly similar results cannot be achieved by closer collaboration at the regional scale by other sectors of the fishing industry. Something akin to the FLAG initiative, modified to suit the circumstances of the UK fishing industry, that encourages fishing enterprises, either collectively or individually, should become a central feature for any future strategy for the development of the coastal fisheries.

Fisheries governance

Thus far we have focused on the need to bring the science and management of the fisheries into the 21st century. There is one further dimension on the restructuring of management to be considered: fisheries governance, a leading theme within this review of the last 50 turbulent years.

When it became clear that the UK was to leave the EU and its remote, monolithic and highly centralised CFP, one of the industry's main hopes was for a form of governance that is more sensitive, interactive and inclusive. It is not clear how far this aspiration can be fulfilled within the current system of devolved governance; discussions over the last four years have focused more on the content of fisheries policy than on the means of delivery. The system is complicated, mainly due to the varying extent to which responsibilities have been devolved to the constituent parts of the UK, with Scotland most fully devolved, and because the UK government at Westminster serves as the governing authority for England's fisheries.

Essentially there are two main concerns for the future. The first is to ensure that the allocation between the three levels of governance – the UK, national and local – is clearly defined, complete and fit for purpose and interaction between the three levels creates a coherent and well-integrated management system for the UK fishing zone as a whole – while allowing some variation to account for regional or local circumstances. Closer to the industry's aspirations for a more interactive and inclusive form of governance, however, is the question of participation in the decision-making process.

The UK has a comparatively good record of actor participation in local fisheries governance: the long-established tradition of co-management in inshore fisheries management – involving local Inshore Fisheries and Conservation Authorities (IFCAs), with devolved powers to regulate fishing activity within the 0–6 nm zone, and the more recent adoption of Regional Inshore Fisheries Groups in Scotland – is a case in point. So too is the redeployment of Fish Producer Organisations (FPOs) as agencies for quota management. But when it comes to major policy decisions, industry participation remains very much at arm's length. Relations between the fishing industry and the government had been improving prior to the Brexit debacle in 2020 but frequent policy-related dialogue between industry leaders and the relevant fisheries department is no guarantee that the industry's point of view will be reflected in the executive decision-making.

What is required is more formal recognition of industry participation at different stages of the policy process. Here, perhaps surprisingly, we can turn to the CFP for evidence of good practice. The Scientific, Technical and Economic Committee for Fisheries (STECF) providing independent advice on policy proposals in their development stages and the regional Advisory Councils offering feedback from an industry perspective on the Commission's policy proposals at later stages of their development, are examples of successful engagement between the governing body and the governed.

Healing the breach: future collaboration between the UK and EU

Drawing artificial boundaries through natural ecosystems for management purposes does not usually make much sense, even less so when the medium is liquid and the populations to be managed highly mobile. Drawing these boundaries through a series of semi-enclosed spaces is, in fact, destructive rather than constructive, fraught with difficulties and at risk of endangering the sustainability of fish populations, should the two management strategies prove incompatible. Such a scenario was not the aim of Brexit, but it will take concerted action on the part of the UK and EU to ensure it does not become the unintended consequence. The ease with which a robust framework for collaboration can be created will depend upon a genuine willingness to work together and activate new negotiating skills that lead to a mutual, constructive and sustainable compromise.

It is to be hoped that collaboration can be extended to include a shared vision and common strategy for the management of the regional seas that have been split apart by Brexit. For this to happen, it will be important for both managers and fisheries organisations to keep open as many channels of communication as possible to avoid developing a culture of isolation. The closing of doors to UK participation in European institutions such as the Advisory Councils is understandable but regrettable. Granting the UK observer status in the case of ACs and the EU's Association of Fish Producer Organisations (EUAPO) would be a welcome step. Continued collaboration should then evolve organically as the realisation of the need for realignment of policy approaches to the realities of managing a shared and vulnerable resource become more apparent.

The bigger picture

Our exploration of the post-Brexit future has so far concentrated on what might be termed 'matters arising', dealing essentially with the fall out from Brexit, the opportunities and challenges it presents and the constraints imposed on the continuing need to work alongside our closest neighbours. Here the options are relatively clear and it is largely a matter of choice as to the direction we intend to take. However, we approach the end of this review of half a century of troubled times for Europe's fisheries with a brief glimpse into a much less certain long-term future, scanning a period that takes us up to the mid-21st century and scaling up our scrutiny from the local to the global scale.

Dominating our images of this future is of course the threat from global climate change. We have grown accustomed to the avalanche of dramatic changes occurring in the high latitudes of both the northern and southern hemispheres that signal the accelerating progress of global climate change. This is at variance with our own experience and perception of climate change as a slow, incremental and scarcely detectable process that has so far done little to trouble the minds of the fishing industry and policy makers unsurprisingly wrapped up in concerns over the present and immediate future. Expert predictions for 2050 suggest that while the tropical seas are likely to experience very serious declines in productivity, temperate waters will be less severely impacted. The bad news is that the shallow waters of continental shores seem certain to experience considerable disturbance. Rising sea levels and warming of the waters will significantly alter marine habitats and ecosystems. Some important cold water species are expected to move offshore into the deeper waters of the continental slope, often beyond the limits of the EEZs. Such processes are already making their mark on fisheries in the waters around Britain.

Rather than dwell too long on how global climate change might impact on fisheries management, we should perhaps consider the bigger picture as to how the main sources of potential instability – environmental, economic and political – might interact to compound the problems of food security and the functioning of the seafood supply chain. As far as we can judge, the importance of fish within our overall food security is guaranteed for the medium and, quite probably, long-term future. Compared with the radical and somewhat confusing scenarios for future agricultural production, sea fishing seems destined to remain a hunting activity reliant on the performance of the natural environment. Provided global warming can be slowed down, with marine ecosystems retaining much of their existing productivity and exploitation levels on fish stocks remaining at or above MSY, it should be possible to avoid a catastrophic downturn in production in North Atlantic waters. These are, however, major provisos.

A more likely source of impact on Europe's food security in general and its seafood supply chains in particular could result from a weakening of the globalised economy and a decline of international trade in fish and fish products. The catalyst for reduced trading is most likely to occur in the tropical belt as developing countries look to protect their own food security, diverting would be exports into the domestic supply chain. The UK's food security strategy is presently highly dependent on trade: we import around 70% of the fish we consume and export a roughly similar share of what we catch. Admittedly, most of the trade is conducted within the less impacted North Atlantic basin. Over the longer term, however, as a result of the greening of the global economy and the quest among developing nations for a greater self-sufficiency in food production, we may have to rely more on consuming what we catch within our own waters than on generating revenue through sales to overseas markets.

How then should we respond to these future challenges? Is it not better to wait until the detailed picture becomes clearer than attempt to address them now? The answer to this question is both 'yes' and 'no'. What we can do is begin the much-needed transformation of fisheries management, from its deeply embedded attachment to 20th-century modes of thinking in both the science and management of fisheries structured around equilibrium models that have led to the creation of simplistic, reiterative management plans, to a way of thinking that embraces uncertainty and the threats of potentially catastrophic disequilibrium. Both marine environmental and fisheries management will need sooner or later to abandon their preservation strategies and replace them with a management approach that builds in the need for greater flexibility, resilience and adaptive responses. There is no better time than the present to start the process that will improve performance now and prepare the industry for the challenges of the future.

If we strike the right balance in our future relationships with the EU and learn the lessons from the past 50 years when developing robust but flexible means of delivering future fisheries policies, then we can address present and long-term challenges with greater confidence. These are big 'ifs', but an alternative future of anarchy, conflict and continued stressing of the marine environment is almost too alarming to contemplate. Failure to contain global warming within the +1.5°C limit by 2050 could, for example, cause profound changes in the global system of ocean currents, including the collapse of the Gulf Stream that underpins the unique diversity and productivity of the seas around the British Isles and provoke a seismic event on a scale that would eclipse the events of the past 50 years.

Further reading

Royal Commission on Environmental Pollution. (2004). *Turning the Tide: Assessing the Impact of Fisheries on the Marine Environment*. London: The Stationery Office.

Scottish Government. (2021). *The Brexit Referendum 5 Years on – Summary of Impacts to Date*. Edinburgh: Scottish Government.

Seafood Task Force. (2021). *Navigating Change: The Way Forward for Our Seafood Sector and Coastal Communities in the Wake of the EU/UK Trade and Cooperation Agreement*. Dublin.

Media sources, notably *Fishing News*.

11

FINAL REFLECTIONS (UNFINISHED)[1]

Introduction

Troubled Waters has focused attention on three major political events – loss of distant waters, the creation and development of the Common Fisheries Policy and Brexit – that have shaped the circumstances of Europe's fisheries and their management over the past 50 years. In focusing on the events and their implications for management, the analysis has perhaps blurred an overall view of how the activity of fishing and fisheries management has changed, and perhaps more importantly how the relationships between the governing system and those who are governed have evolved. These final reflections should go some way towards restoring the imbalance.

The catching sector

The changes to the catching sector over the past 50 years are striking. Europe's fishing fleet today is very much smaller in terms of the number of vessels and employment. However, the average vessel is larger in size, engine power and fishing capacity. In vessel design, fishing gears and onboard technology, the modern fishing enterprise is much more sophisticated with significant innovations ranging from fish detection, the design and handling of fishing gear, onboard storage of the catch to communications between ship and shore. Such improvements enable the skipper to go to sea with a safer, faster and more efficient vessel and a smaller crew. Modernisation of the fishing enterprise is visible in virtually all sectors and segments of the fleet but most strikingly in the pelagic sector where a myriad of small boats have been replaced by a handful of very large vessels that come closest to the image of an industrialised fishing industry. Despite contraction in the size of the fleet, structural change is limited: Europe's fishing industry still comprises

DOI: 10.4324/9781003362913-14

a very large and highly diverse number of relatively small independent fishing enterprises. And in one important sense, the economic modernisation of the fleet is incomplete: what is missing is the widespread adoption of company ownership. It is the diversity, dispersal and fragmented ownership of the catching sector that makes the fishing industry difficult to manage.

Fisheries management

In many ways, the transformation of fisheries management over the past 50 years has been even more spectacular. In 1970 Europe's fishing fleets enjoyed the freedom of the high seas embedded in the international law of the sea. There was simply no political framework within which to develop a regulatory system for managing the fisheries. The only areas where the coastal state could exert its authority were the very narrow territorial waters extending out to a maximum of ten or twelve nautical miles and treated as the state's exclusive fishing zone. Surprisingly few European countries had seized the opportunity to create integrated management systems for their inshore fisheries: the UK, or more accurately England and Wales, was the major exception, initiating local co-management organisations as early as the 1880s.

Following the implementation of EEZs and thus the curtailment of the freedom of the high seas in the mid-1970s, the majority of coastal states were quick to respond to the newly conferred responsibilities for managing the fisheries. Within a matter of a few years, most had established basic rules for conserving the stocks and limiting access to foreign vessels. Previous experience in fisheries management was non-existent: the necessary skills had to be learned on the job. Likewise, monitoring and enforcement capabilities had to be greatly expanded. Over the next 30 years, these basic arrangements were translated into more comprehensive, stricter management regimes in an attempt to arrest the persistent decline in several key fish stocks.

The term fisheries policy tends to refer to a narrowly constructed, science-based set of actions designed to aid the conservation of commercial fish stocks. For those with responsibility for implementing policy, there is a limited range of measures in their toolkits. Practically all North Atlantic states have opted for a system of output control in the form of catch quota closely aligned to MSY to contain the impact of fishing on the stocks. This is complemented by a range of technical conservation measures (TCMs) that include gear regulations, minimum landing size for individual species and closed areas/seasons intended to protect spawning and juvenile stocks.

Of these TCMs, gear regulations are most likely to cause consternation and frustration within the industry. The problem here is one of conflicting intentions. Changes in gear technology are normally intended to improve selectivity in terms of the fish caught, reducing the numbers of undersized juveniles and non-target species, thereby increasing efficiency. On the other hand, management interventions are mostly directed towards limiting the number, size and capacity of the gears being deployed by each vessel; this is likely to offset the benefits of improved technology.

The list of management measures referred to above is incomplete: it represents the basic toolkit found among the majority of Atlantic coastal states. In the event of crisis, such as that occurring around the turn of the century within EU waters when a range of major fish stocks were experiencing severe decline, other more expensive or invasive measures were deployed for a limited period, alongside the established management procedures, including a decommissioning scheme and days at sea restrictions. Decommissioning was voluntary and intended to involve older, less efficient vessels, thus limiting its effectiveness in reducing overcapacity. Days at sea restrictions – a form of input control – were introduced mainly for demersal fishing activity, requiring the vessel to limit its fishing activity to a specified number of days within each month. This proved more effective but for the skipper owner almost the last straw: not only were they being told what, where, how and how much fish to catch but also when they were allowed to do it!

In order to assess the true nature of a management system we need to understand more about its institutional structure, style of delivery and its interaction with the client population – the fisher. Broadly speaking, fisheries management in the North Atlantic falls within the category of a science-led, command and control system. However, this simple description is insufficient when it comes to the more complex situation occurring within the EU that adds a further tier of decision-making where the unelected Commission has exclusive responsibility for initiating new conservation policy measures relating to the setting of annual TACs and quota as well as the deployment of TCMs, subject to the co-approval by the Council of Ministers and the European Parliament. In the initiation of new policy, the Commission is guided by its own highly experienced and resourceful fisheries directorate (DG Mare). Member states, on the other hand, are responsible both for the implementation and monitoring of EU regulations and for certain reserved areas of management including fleet management (vessel licensing), quota management and inshore fisheries.

As for the means of delivering EU policy, the process is simple, direct and unyielding. In fisheries, new or revised policy measures are communicated in the form of Regulations that must be translated directly into member state law, rather than through the more widely used Directives that allow member states some discretion in the way in which a policy decision is integrated within its own law code.

Where the EU's governing system seems to be at its weakest, at least in the eyes of the fishing industry, is in the apparent lack of consultation with the principal stakeholders. Over the past 50 years, relations between the industry and the member states' administrations have grown closer and proved more fruitful. On the industry's part, the representative organisations have become more politically aware and more professional to the point where their opinions and advice are respected and valued by those responsible for fisheries management. In relation to the Commission's leading role in determining the course of fisheries policy, consultation with member states conducted largely through COREPER (Committee of Permanent Representatives i.e. attachés), with scientific, technical and economic experts (STECF) and with the regional Advisory Councils over new policy proposals, appear to function effectively. However, when it comes to consultation directly with those in the fishing industry – essential to building confidence for the management to work smoothly – there is something of a hiatus.

In fact, an Advisory Committee for Fisheries has been in place since the early 1970s, reembedded in the CFP in 1983 and subsequently expanded as the Advisory Committee for Fisheries and Aquaculture in 1999. Somewhat surprisingly, it was discontinued in 2013 leaving many in the industry feeling disenfranchised. ACFA, so the argument went, had lost its purpose. Its consultative role had been subsumed by a range of EU-wide stakeholder organisations, but most importantly by the regional and sectoral Advisory Councils initially set up in 2004, that provided clearer, more focused advice on specific policy issues albeit conditioned by a strong environmental perspective. All of which was no doubt true, but it still left many in the industry all the more convinced that the Commission was both inaccessible and unaccountable and that as the principal stakeholder they remained very much the object rather than the subject of fisheries policy.

The family-based fishing enterprise and the role of the skipper owner

We come finally to consider the changing roles, status and lifestyles of the fisher – or more specifically the skipper owner, the linchpin of the family-based fishing enterprises that continue to dominate the scene in western Europe. Some 50 years ago, small ports and harbours throughout much of the region were bristling with fishing activity, mainly associated with local coastal fisheries that offered a diversity of species and accounted for the greater part of domestic fish production. The coastal fishing vessel was typically an all-purpose boat ranging in size from 12 to 16 m overall length, sometimes extending to 18 or 19 m, and suitable either as dayboats or for making longer trips of two or three days and occasionally longer and capable of accommodating a combination of two or three seasonal fisheries – each requiring its own gear and varying complements of crew members – that together provided for full-time fishing. Out of season, these boats could also be put to use as freight or livestock carriers, and of course in ports that also catered for holiday makers, used as charter vessels for sea anglers, offering visitors 'trips around the bay' or giving them a taste of the fishing experience, 'jigging' for tomorrow's breakfast.

In much the same way that coastal ecosystems were able to meet the industry's requirement in terms of a flexible resource base, so too three key rural social institutions – household, extended family and community – were in a position to satisfy the organisational needs of the family-based fishing enterprise and most importantly provide a full complement of crew. Barring misadventure, the two or three generation family household could provide two or more crew members, with the eldest son on leaving school joining the crew as a deckie learner; other sons might also follow. The enlistment of the eldest son was usually sufficient to secure the continuity of the family enterprise for yet another generation. Additional crew members were likely to be recruited from the extended family living within the same or neighbouring community. Other adult members of the household made up the shore crew, with the skipper's wife taking on the roles of secretary and accountant, ordering supplies for the next trip, while members of the older, retired generation would help in mending the nets, baiting lines and preparing the vessel for its next trip.

Central to the conduct of the family fishing enterprise is the skipper-owner – the sole decision maker when at sea, on whom the success of the enterprise ultimately rests. Within reason, he was a free agent still largely untrammelled by extensive codes of legislation, the vessel, health and safety; he was free to choose when, where, by what means and what species to target. He must also make the crucial decision as to when to suspend fishing and head for the market. His decisions were based essentially on a combination of experience, local ecological knowledge – not simply concerning the location and geography of the available fishing grounds but also the likely behaviour of the fish under particular conditions – familiarity with his vessel and gears and the local market. Crucially he needed to know his crew, their individual strengths and weaknesses, limits of endurance and their ability to work together as a team. Each trip involved a risk assessment concerning the state of the stocks, tidal and weather conditions and the likely state of the market. A successful skipper is more likely to be a risk taker than risk averse.

Alongside the apparent freedom of choice, the skipper must also shoulder several responsibilities, not only regarding the success of the individual trip and the performance of the enterprise over the year and the welfare of his own family but also the health and safety of his crew while at sea and providing a fair reward for their work. Coastal fishing in the 1970s usually involved 'share fishing' where instead of receiving a basic wage, crew members' income is determined by the overall profit of the trip. Once the operating costs have been accounted for, the profits are shared between the vessel, the skipper and the crew according to their seniority and experience.

Success was measured mainly by the size of the catch, rewarded not only by reliable income but also by the support and loyalty of the crew, guaranteeing the continuity of the enterprise, and widely acknowledged throughout the fishing community as a whole. Top skippers may eventually become part of the folklore of the community. Coastal fishing is a tough, physically demanding and unforgiving occupation often undertaken in hostile conditions. It is also highly competitive; at sea, neighbouring enterprises from within the same community are rivals, rarely coming together to share good fortune but tending rather to protect their interests through remaining silent and even allowing the spread of false information!

Fifty years on, the nature of coastal fishing has changed appreciably. Though the fleet is smaller in numbers, the individual vessel has become more specialised in design, with improved facilities for the crew and for processing and storage of the catch, increased engine capacity that reduces the time travelling to and from the grounds and more operationally efficient through investment in labour-saving technology such as power hauling of the gear. The family-based enterprise has become more economically aware: this is reflected in the switch from size to quality of the catch – and the price it earns on the quayside – as the indicator of a successful trip and further encouraged by the expansion in coastal waters of high-value fin- and shellfish fisheries destined mainly for major urban markets in continental Europe.

However, little of the seemingly stable, coherent and harmonious image of the social structure of coastal fishing has remained intact. The inherent remoteness of many coastal communities and the gradual urbanisation of the fishing industry for all but inshore and smaller coastal enterprises, together with the changing economic relationships between the industry and both financial and market institutions, can all be cited as contributing to the weakening of the traditional social structures.

There are certain instances especially in more highly urbanised coastal states where, despite proximity to reliable fishing grounds, the industry struggles for survival due to external pressures of urbanisation.[2] In some smaller fishing settlements retirement migration to the coast, coupled with second home ownership, which expanded considerably over the past 50 years have put tremendous pressure on limited housing stock and effectively priced the fishing community and particularly the younger generation out of the local housing market. Such adventitious elements of the population where there are sufficient numbers may come together to oppose future developments in housing or commercial property.

The second and potentially more damaging source of pressure comes with the growth of tourism and leisure industries and in the creation of the holiday resort. The latter is not a recent phenomenon; in Britain it dates back to the mid-19th-century expansion of the railway network connecting larger coastal settlements with an urban hinterland, enabling day trips to the seaside, and the somewhat later provision of paid annual leave allowing for longer visits. Competition increased not only for property to accommodate the visitor but also prime quayside estate for the development of tourism or visitor-related services and also for space within the harbour itself for pleasure craft and privately owned yachts and motor cruisers. Physically, economically and culturally the townscapes of many resorts have been completely transformed. Faced with an alien, if not hostile, seafront that is bold, brassy and brazen in season and deserted and tawdry out of season, what remains of the fishing community is at best isolated and at worst abandoned where fishing is respected as heritage rather than as a viable component of the local economy.

But by far the most important cause of a weakening of traditional social structures is to be found in the greatly altered social dynamics of the fisher household, family and community that vary in strength throughout Europe. Due to processes of demographic transition, the fisher household has become simpler in structure and smaller in size, bringing it closer to the two generation, nuclear household, with fewer offspring characteristic of modern urban society. Extension of school leaving age, provision of tertiary education facilities and, in some areas, a broadening of the local labour market have all conspired to reduce the likelihood of children opting for employment within the family enterprise as the crucial first step towards securing its succession and continuity. In effect, fishing is no longer an occupation of necessity or obligation but one of choice involving hard physical graft, unsocial hours and variable incomes. It is less appealing to the younger generation. These same basic circumstances also apply to the extended family – fewer in number and dispersed over a wider geographical area, it too is less likely to prove a reliable source of recruitment for the family fishing enterprise. While the traditional family-based enterprise remains the preferred option, providing the best guarantees for success and succession, it has become much harder to achieve.

With the local community repeating the pattern of fewer active households and a younger generation able to choose from a much wider range of occupational options, the situation facing many family-based enterprises has become quite critical. Increased turnover and dependence on non-local and indeed non-national crew members to secure the viability of the enterprise have meant a significant change to the ethos of the enterprise at sea and the skippers' perception of reliability based on familiarity with both kith and kin. The changing ethos has been reinforced by the replacement of 'share fishing' by fixed wage rates as the basis for remuneration. Scaling down the required complement of crew has become both plausible and inevitable due to the trickle-down process of technical, labour-saving innovations such that small coastal vessels can reduce their crew numbers from six to four and the inshore sector from two or three to a lonely single-handed operation.

There can be no doubt that the life and livelihoods of the skipper owner have been greatly altered by these changing circumstances and his role and status somewhat diminished as a result of modern fisheries management. What in the early 1970s was largely the skipper's call was denied him through the combination of restrictive licensing, quota entitlements and the introduction of certain technical conservation measures. But his status within the local community was weakened not so much by these constraints as by the dilution of the so-called fishing community itself. By the early 2020s, there were usually fewer households with a direct involvement in fishing and fewer households with sufficient knowledge and understanding of, or interest in, fishing to be able to discern the qualities that contributed to the skipper's status.

For many the job remains basically the same. The skipper retains his responsibilities in respect of the vessel, recruitment of crew and their health and safety on board and above all the wellbeing of the family household. He is still required to decide, within certain constraints, when and where to fish, the preferred composition and presentation of the catch and after dialogue with his agent the optimal timing of its delivery to the quayside market. Among the older skippers with no clear line of succession, there may well be the unenviable task of when to call a halt to proceedings, putting the vessel up for sale, often with its most valuable asset – the quota entitlement – being handled separately. It is not uncommon for it to serve as a nest egg to support the older generation in retirement through annual leasing and earning the former skipper the tag of 'slipper skipper'. There is little or no compensation for the unemployed crew.

A small but growing number of younger skipper owners are taking a more entrepreneurial approach to their fishing enterprise, seeking to maximise the value of the catch often through direct sales or by adding value through processing. Acting either individually or collectively, they may target local gastro-tourism businesses, establish local delivery networks focusing on 'catch of the day' consignments or lease stalls at farmers' markets and/or weekly outdoor urban markets. Alternatively, they may look to more ambitious and expensive value-adding processing to create specialist dishes using species for which the region has a particular reputation. Not only do such ventures add value, they also provide additional employment opportunities for family members or the wider community and so help to strengthen the resilience of not only the individual enterprise but also the local community and the fisheries sector in general.

Author's jottings and unincorporated notes accompanying Chapter 11 manuscript[3]

Note 11.1

Need to find space for comments on:

1. Chapter 6 overview of CFP. Throughout the time working on CFP I held a broadly optimistic view that once those responsible for formulating/operating the CFP had developed greater confidence ... achieved maturity ... the Policy would become more flexible and better adapted to differing situations ... [The] 2012 decennial Review marking the 30th anniversary of the Policy was an appropriate time for that confidence and maturity to become apparent; [the] Commission's failure to respond positively to DG Mare's own proposals for a more thorough reform was the last straw. Left to wonder whether the Policy will gradually become more archaic and less effective (CFP has given us the real benefits of a stable situation regarding [the] EU's fisheries sector ...)
2. Comment on Chapter 10 regarding the considerable shock of UK departure from EU, puncturing the sense of permanence, stability ... & ... comment on EU's mature performance during 2020 negotiations and later. Its quiet authority in damping down France's demands for legal action v UK [over fishing licences].

Note 11.2

The success of a fisheries policy is defined as much by the things it avoids doing, as by the actions it takes

- Failure of the EU's conservation policy in the past is largely attributable to reluctance to engage with the issue of *discards* (allowing fishers to catch far more fish than their quotas ... fishers could 'correct' the problem by discarding illegal catch and also resort to high grading of the catch intended for the market...), *mixed demersal fisheries* and the issue of *latent capacity* (many sectors have catching capacity far in excess of [what] the quota requires either within the active fleet *or* in vessels temporarily or permanently out of action)
- Emergency measures (e.g. those introduced temporarily around the turn of the century to reduce capacity) = decommissioning schemes; introduction of input control measures (days at sea), temporary closed areas ... Of these days at sea = most feared / bitterly resented. Completed the ... control over fishing activity. i.e. where (licensing, closed areas), what (licensing and quota), how (gear regulations), how much (quota) and when (days at sea restrictions) to fish – completing the management's control over fishing activity [and a] challenge to the identity of status of the skipper.

Note 11.3

Need to rethink the Conclusion:
Partly to summarise the impact of the past 50 years *viz*:

- lessons from the past that go unlearned
- cannot simply ignore the course of history and its legacy for the future
  - 1970s – [UK] unwittingly (?) opted for membership of EEC rather than independence – had it been the other way round might well have been a very different story
  - agreed to perpetuate the concept of freedom of the high seas in respect of the common pond and for a club of seven fishing nations, that was over the years to [expand its] members, as opposed to claiming the full sovereign rights (and responsibilities) to the UK's own 200 EEZ; and later affirmed the full implications of that move in the loaded negotiations leading to the CFP in 82.
  - we had inadvertently underwritten the expectations of neighbouring states like Belgium and the Netherlands to rely on fisheries within the UK's segment of the common pond for almost 1/3 of their allowed catch.

- consciously or otherwise we were placing the wider economic benefits of membership ahead of the narrower interests of the fishing industry …, only to repeat the 'mistake' some 40 years later on when 'negotiating' Brexit.
- cannot wind the clock back to the early 1970s: obliged to accept the consequences of the last 50 years or so and the fact that the European Commission had grown in stature as a negotiating power to become a formidable opponent and stalwart protector of its members' interests.
- very unlikely to gain much from any further renegotiation in 2025 – hard to imagine countries like France or the Netherlands agreeing to give up their entitlements to fish in the 6–12 nm zone of the UK's territorial waters; the likely price demanded for such concessions would be too much, namely the loss of tariff-free access to the EU's markets for the UK.
- much more beneficial to the UK industry's long-term interests is to accelerate progress towards close collaboration with the EU over an agreed approach to manage resources/exploitation within the shared waters that surround much of the UK
- on a very different scale, the achievements of the FLAG scheme have clearly pointed to the opportunities of forging closer relations between fishing and the broader local economy, ecosystem and environmental concerns
- What we call fisheries policy is largely confined in its aims and actions, to managing fish stocks per se; attention paid to other relevant issues especially social justice, food security, ecosystem sustainability is usually quite minimal despite claims made in policy statements; so too is the growing influence of [the] market-based approach that allows for sale/leasing of fishing rights as a mechanism for removal of excess fishing capacity and the essential balancing of available resources and fishing capacity.

Note 11.4

FUTURE

Fishers are rarely trusted to manage the resources themselves cf traditional fisheries – some understanding that by working together within a broad set of guidelines can have an effect

- competitiveness of the skipper not diminished – his independence.
- few successful attempts by governing system to give fishers more say/ greater role in their management.

- co-management systems assume a well-developed set of limits within which the fishers (or those that represent their interests) are given the scope to manage.
- plenty of examples at the local scale; some examples regarding a limited set of responsibilities at national scale.
- few signs that fishers are seeking to desert their independence, individuality, in favour of closer collaboration (self-managing) schemes e.g. unwillingness to take on community quota management schemes for small-scale sector.

Note 11.5

Decide what still has to be included:

1. Viz skipper owner
2. Omission regarding the success of policy [in terms of] what is there/not there
3. Interactive governance (not yet developed)
4. Prospects for the future (see separate sheet)

Etc.

Notes

1 Manuscript of chapter unfinished at time of author's death.
2 This paragraph and the next, which accompanied the handwritten manuscript, have been located in the text by Jeremy Phillipson.
3 Edited and presented by Jeremy Phillipson.

INDEX

Note: *italic* page numbers refer to figures and page numbers followed by "n" denote endnotes.